Table of Contents

Introduction

Chapter 1: Garage Door Springs
- Types of Garage Door Springs
- Signs of Worn-Out or Broken Springs
- How to Measure and Order Replacement Springs
- Step-by-Step Guide to Replacing Torsion Springs
- Step-by-Step Guide to Replacing Extension Springs
- Adjusting Spring Tension for Optimal Performance

Chapter 2: Garage Door Openers
- Types of Garage Door Openers
- Choosing the Right Opener for Your Needs
- Installing a New Garage Door Opener
- Programming and Adjusting Your Opener
- Troubleshooting Common Opener Issues

Chapter 3: Garage Door Tracks
- Understanding Garage Door Track Components
- Identifying Misaligned or Damaged Tracks
- Aligning and Adjusting Horizontal Tracks
- Aligning and Adjusting Vertical Tracks
- Lubricating and Maintaining Tracks for Smooth Operation

Chapter 4: Garage Door Panels
- Assessing Panel Damage
- Repairing Minor Dents and Cracks
- Replacing Individual Panels
- Painting and Weatherproofing Panels
- Insulating Garage Door Panels for Energy Efficiency

Chapter 5: Additional Maintenance and Troubleshooting
  - Maintaining and Replacing Garage Door Rollers
  - Adjusting and Replacing Garage Door Cables
  - Troubleshooting Noisy Garage Doors
  - Fixing a Garage Door That Won't Open or Close
  - Weatherstripping and Sealing Your Garage Door

Conclusion

Introduction

Picture this: you're running late for work, and as you frantically hit the button on your garage door opener, you hear a loud, unsettling snap. Your garage door refuses to budge, and you're left staring at a broken spring, a misaligned track, or a damaged panel. Frustration sets in as you realize you're stuck, unable to get your car out of the garage. Sound familiar?

As a homeowner, your garage door is one of the most essential and frequently used components of your home. It provides security, convenience, and protection for your vehicles and belongings. However, like any mechanical system, garage doors are prone to wear and tear, and when neglected, they can cause significant inconvenience and even pose safety risks.

That's where "How to Repair and Maintain Garage Doors" comes in. This comprehensive guide is designed to empower you with the knowledge and skills necessary to tackle common garage door problems head-on. Whether you're dealing with a broken spring, a malfunctioning opener, a misaligned track, or a damaged panel, this book will walk you through the step-by-step process of diagnosing and fixing these issues like a pro.

But why should you bother learning how to repair and maintain your garage door yourself? Well, for starters, professional garage door repairs can be expensive. By mastering the skills outlined in this guide, you can save hundreds, if not thousands, of dollars in repair costs over the lifetime of your garage door. Plus, you'll have the satisfaction of knowing that you can handle any garage door problem that comes your way, without having to wait for a technician to come to your rescue.

In this book, we'll cover everything from the most common garage door problems and their causes to the essential safety precautions and tools you'll need to get the job done right. We'll demystify the inner workings of your garage door, helping you understand the function and importance of each component, so you can diagnose issues with confidence.

But don't worry – we won't bog you down with technical jargon or complicated theories. Our goal is to make garage door repair and maintenance accessible to everyone, regardless of their skill level or experience. With clear, concise instructions, helpful illustrations, and real-life anecdotes, you'll feel like you have a trusted friend guiding you every step of the way.

So, whether you're a DIY enthusiast looking to expand your home maintenance repertoire, or simply a homeowner who wants to be prepared for any garage door emergency, "How to Repair and Maintain Garage Doors" is the ultimate resource. By investing in this guide, you're not just buying a book – you're investing in the peace of mind that comes with knowing you can keep your garage door running smoothly and safely for years to come.

Are you ready to take control of your garage door and become your own garage door hero? Let's dive in and get started on this empowering journey together!

Chapter 1
Garage Door Springs
Types of Garage Door Springs

Garage door springs are the unsung heroes of your garage door system. These powerful components are responsible for counterbalancing the weight of your garage door, making it possible to lift and lower the door with ease. Without properly functioning springs, your garage door would be incredibly heavy and nearly impossible to open manually.

In this chapter, we'll dive deep into the world of garage door springs, exploring the different types available and their unique characteristics. By understanding the role springs play in your garage door's operation, you'll be better equipped to identify issues and make informed decisions when it comes to repairs and replacements.

Types of Garage Door Springs

There are two main types of garage door springs: torsion springs and extension springs. Each type has its own set of advantages and disadvantages, and the choice between them often depends on factors such as your garage door's size, weight, and design.

1. Torsion Springs

Torsion springs are the most common type of garage door springs found in modern homes. These springs are located above the garage door opening, parallel to the horizontal tracks. They are mounted on a metal shaft and secured to a center plate, which is attached to the top of the garage door.

How Torsion Springs Work:

When the garage door is closed, the torsion springs are tightly wound, storing a significant amount of energy. As you begin to open the door, the springs unwind, releasing their stored energy and counterbalancing the door's weight. This process makes lifting the garage door much easier, as the springs bear most of the weight.

Advantages of Torsion Springs:
- Durability: Torsion springs typically last longer than extension springs, with an average lifespan of 15,000 to 20,000 cycles (one cycle equals one full opening and closing of the door).
- Balanced lifting: Torsion springs distribute the door's weight evenly, providing a smoother and more stable operation.
- Space-saving: Since torsion springs are mounted above the door opening, they don't require additional side room, making them ideal for garages with limited space.
- Safety: If a torsion spring breaks, it remains contained on the shaft, reducing the risk of injury from flying parts.

Disadvantages of Torsion Springs:
- Complexity: Installing and adjusting torsion springs can be more complex than extension springs, often requiring professional assistance.
- Cost: Torsion springs are generally more expensive than extension springs, both in terms of the parts themselves and the cost of professional installation.

2. Extension Springs

Extension springs are an older style of garage door spring, but they are still found in many homes, particularly those with older garage door systems. These springs are located on either side of the garage door, running parallel to the horizontal tracks.

How Extension Springs Work:
Extension springs are stretched when the garage door is closed, storing energy in their extended state. As the door opens, the springs contract, releasing their stored energy and helping to lift the door. The springs are attached to cables that run through pulleys, connecting the springs to the bottom corners of the garage door.

Advantages of Extension Springs:
- Cost-effective: Extension springs are typically less expensive than torsion springs, making them a more budget-friendly option for homeowners.
- Easier installation: Installing extension springs is generally less complex than installing torsion springs, making it a more manageable task for DIY enthusiasts.

Disadvantages of Extension Springs:
- Shorter lifespan: Extension springs typically have a shorter lifespan than torsion springs, lasting around 10,000 cycles on average.
- Less balanced lifting: Extension springs may not distribute the door's weight as evenly as torsion springs, potentially leading to a more uneven or jerky operation.
- Space requirements: Extension springs require additional side room, as they are mounted on either side of the garage door. This can be problematic for garages with limited space.
- Safety concerns: If an extension spring breaks, it can fly off with significant force, posing a safety risk to anyone nearby. Safety cables are essential to help contain broken springs and prevent injuries.

Choosing the Right Springs for Your Garage Door
When selecting replacement springs for your garage door, it's crucial to choose springs that are compatible with your door's size, weight, and design. Improperly sized springs can lead to poor performance, premature wear, and even safety hazards.

Factors to consider when choosing garage door springs include:

1. Door weight: The weight of your garage door determines the amount of tension needed in the springs to counterbalance the door effectively.
2. Door size: The width and height of your garage door influence the size and number of springs required.
3. Spring cycle life: Consider the expected lifespan of the springs and choose a cycle life that matches your usage needs and budget.
4. Material quality: Opt for high-quality spring materials, such as oil-tempered steel, to ensure durability and longevity.

In most cases, it's best to consult with a professional garage door technician to help you select the appropriate springs for your specific garage door. They can assess your door's requirements and recommend the best options based on your needs and budget.

Conclusion

Garage door springs are a critical component of your garage door system, responsible for counterbalancing the door's weight and enabling smooth, safe operation. By understanding the different types of springs available and their unique characteristics, you can make informed decisions when it comes to maintaining, repairing, or replacing your garage door springs.

Remember, working with garage door springs can be dangerous due to the high tension they are under. If you're unsure about your ability to safely handle spring repairs or replacements, it's always best to seek the assistance of a qualified garage door professional.

In the following chapters, we'll explore the signs of worn-out or broken springs, how to measure and order replacement springs, and provide step-by-step guides for replacing both torsion and extension springs. Armed with this knowledge, you'll be well on your way to becoming a garage door spring expert and ensuring the longevity and reliability of your garage door system.

Signs of Worn-Out or Broken Springs

As a homeowner, it's essential to be aware of the signs that indicate your garage door springs are wearing out or have broken. By recognizing these signs early on, you can address the issue before it leads to more severe problems or safety hazards. In this section, we'll explore the most common signs of worn-out or broken garage door springs and what they mean for your garage door system.

1. Difficulty Opening or Closing the Garage Door

One of the most obvious signs of worn-out or broken springs is difficulty in opening or closing your garage door. If you notice that your garage door feels heavier than usual or requires more effort to lift, it's likely that your springs are losing their tension and are no longer able to counterbalance the door's weight effectively.

This issue can manifest in several ways:
- Manual operation: If you have to exert significantly more force to lift your garage door manually, it's a clear indication that your springs are weakening.
- Automatic operation: If your garage door opener appears to struggle when lifting the door, making unusual noises, or reversing unexpectedly, worn-out springs may be the culprit.

2. Uneven Garage Door Alignment

When your garage door springs are wearing out or have broken, you may notice that your garage door appears uneven or crooked when it's partially open. This is particularly common with extension spring systems, where one spring may wear out faster than the other, causing an imbalance in the door's alignment.

An uneven garage door can lead to several problems:
- Increased wear on door components: An unbalanced door puts additional stress on the tracks, rollers, and hinges, leading to premature wear and tear.
- Reduced energy efficiency: Gaps created by an uneven door can allow drafts and pests to enter your garage, compromising energy efficiency and home comfort.
- Safety hazards: An unbalanced door may not close properly, posing a security risk and increasing the likelihood of accidents or injuries.

3. Visible Signs of Wear or Damage

Visually inspecting your garage door springs can reveal signs of wear or damage that indicate the need for replacement. Some common visible signs include:

- Rust or corrosion: Over time, exposure to moisture and humidity can cause springs to rust or corrode, weakening their structural integrity and increasing the risk of failure.
- Cracks or gaps: Small cracks or gaps in the springs' coils can be an early warning sign of impending failure. As these cracks worsen, the springs become more likely to break suddenly.
- Stretching or elongation: Extension springs that appear stretched out or longer than their original length have likely lost their tension and are no longer able to support the door's weight effectively.

4. Unusual Noises During Operation

If you notice unusual noises coming from your garage door during operation, it could be a sign that your springs are wearing out or have broken. Some common noises to listen for include:

- Squeaking or grinding: These noises may indicate that your springs are rubbing against other components due to improper alignment or lack of lubrication.
- Popping or banging: Loud popping or banging sounds, particularly when the door is opening or closing, can be a sign that your springs have broken or are about to break.

5. Age of the Garage Door Springs

Even if you haven't noticed any of the above signs, it's important to consider the age of your garage door springs. Most torsion springs are rated for around 10,000 cycles (one cycle equals one full opening and closing of the door), while extension springs typically last for about 10,000 cycles.

If your garage door is used frequently, your springs may wear out sooner than expected. As a general rule, if your garage door is more than 7-10 years old and has never had its springs replaced, it's a good idea to have them inspected by a professional to assess their condition and determine if replacement is necessary.

Conclusion

Recognizing the signs of worn-out or broken garage door springs is crucial for maintaining the safety, efficiency, and reliability of your garage door system. By being aware of these signs and addressing them promptly, you can prevent more serious issues from developing and ensure that your garage door continues to operate smoothly.

If you suspect that your garage door springs are wearing out or have broken, it's essential to exercise caution and avoid attempting to operate the door until the issue has been resolved. Garage door springs are under high tension and can cause serious injury if handled improperly.

In the next section, we'll discuss how to measure and order replacement springs, ensuring that you have the correct parts on hand when it's time to tackle the repair process. Remember, if you're ever unsure about your ability to safely replace your garage door springs, don't hesitate to contact a professional garage door technician for assistance.

How to Measure and Order Replacement Springs

Measuring and ordering the correct replacement springs for your garage door is crucial to ensure proper function, safety, and longevity. Incorrect spring sizes or types can lead to poor performance, premature wear, and even dangerous situations. In this section, we'll walk you through the process of measuring your existing springs and ordering the appropriate replacements.

Step 1: Identify the Type of Springs

Before measuring your springs, you must first identify whether your garage door uses torsion springs or extension springs. This information is essential, as the measuring process differs for each type.

- Torsion springs: Located above the garage door opening, parallel to the horizontal tracks. They are mounted on a metal shaft and wind or unwind to assist in lifting and lowering the door.
- Extension springs: Located on either side of the garage door, running parallel to the horizontal tracks. They stretch and contract to counterbalance the door's weight.

Step 2: Gather the Necessary Tools

To measure your garage door springs accurately, you'll need the following tools:

- Measuring tape
- Marker or pencil
- Notepad
- Safety goggles and gloves (for protection)

Step 3: Measure the Springs

A. Torsion Springs:
1. Starting at one end of the spring, measure the length of the spring from end to end, including any winding cones or other hardware. Record this measurement in inches.
2. Measure the inside diameter of the spring wire. To do this, use a tape measure or caliper to determine the distance between the inside edges of the spring coils. Common sizes include 1.75", 2", and 2.25".
3. Count the number of coils in the spring. This can be done by marking a starting point on the spring with a marker, then counting each coil until you reach the starting point again.
4. Measure the length of the winding cone, if applicable. This is the portion of the spring that attaches to the winding bracket.

B. Extension Springs:
1. Measure the length of the spring from eye to eye, including any hooks or loops. Record this measurement in inches.
2. Measure the width of the spring from the outside edges of the coils. Record this measurement in inches.
3. Determine the wire size of the spring. This can be done using a wire gauge or by measuring the diameter of an individual coil with a caliper.
4. Check for any color coding or labeling on the springs, as this information can be helpful when ordering replacements.

Step 4: Determine the Wind Orientation (Torsion Springs Only)

For torsion springs, you'll need to determine the wind orientation. This refers to the direction in which the spring is wound and is essential for ensuring proper installation and function.

- Left-wound springs: When viewed from the left side of the door (from inside the garage), a left-wound spring will have coils that appear to go "up" from the winding cone.
- Right-wound springs: When viewed from the right side of the door (from inside the garage), a right-wound spring will have coils that appear to go "up" from the winding cone.

Step 5: Order Replacement Springs

With your measurements and spring type information in hand, you can now order replacement springs from a garage door supplier or manufacturer. When placing your order, be sure to provide the following information:

- Type of spring (torsion or extension)
- Length of the spring
- Inside diameter of the spring wire (torsion) or width of the spring (extension)
- Number of coils (torsion) or wire size (extension)
- Wind orientation (torsion only)
- Any additional measurements, such as winding cone length or color coding

It's essential to order springs that are identical to your existing springs in terms of size, type, and load capacity. If you're unsure about any aspect of your measurements or the ordering process, consult with a professional garage door technician to ensure you receive the correct replacement springs.

Step 6: Safety Considerations

When measuring and handling garage door springs, always prioritize safety. Garage door springs are under high tension and can cause serious injury if mishandled. Always wear safety goggles and gloves when working with springs, and never attempt to adjust or release the tension on a spring without proper training and tools.

If you're not confident in your ability to safely measure or replace your garage door springs, it's best to contact a professional garage door technician for assistance.

Conclusion

Accurately measuring and ordering replacement garage door springs is a critical step in maintaining the proper function and safety of your garage door system. By following the steps outlined above and providing precise measurements to your supplier, you can ensure that you receive the correct replacement springs for your specific garage door.

Remember, if you're ever unsure about any aspect of the measurement or ordering process, or if you're not comfortable handling garage door springs on your own, don't hesitate to seek the assistance of a qualified garage door professional. They can help you identify the correct replacement springs and ensure a safe and proper installation.

In the next section, we'll provide a step-by-step guide to replacing torsion springs, walking you through the process from start to finish. With the right knowledge, tools, and safety precautions, you can successfully replace your garage door springs and restore your door to optimal functionality.

Step-by-Step Guide to Replacing Torsion Springs

Replacing torsion springs on your garage door can be a challenging and potentially dangerous task due to the high tension under which these springs operate. However, with the right tools, knowledge, and safety precautions, you can successfully replace your torsion springs and restore your garage door to proper functioning. In this section, we'll provide a detailed, step-by-step guide to replacing torsion springs.

Safety Warning: Torsion springs are under extreme tension and can cause serious injury or death if mishandled. If you're not confident in your ability to safely replace your torsion springs, it's strongly recommended that you contact a professional garage door technician for assistance.

Tools and Materials Needed:
- New torsion springs (properly sized for your door)
- Winding bars (two pieces of solid steel rods, typically 18" long)
- Adjustable wrench or socket set
- Vise grips
- Ladder
- Safety goggles and gloves
- Tape measure
- Marker or pencil
- Hammer
- Flathead screwdriver
- Replacement center bearing (if necessary)

Step 1: Prepare the Garage Door
1. Close the garage door and unplug the garage door opener (if applicable) to prevent accidental operation during the repair process.

2. Place a C-clamp or locking pliers on the track just above one of the rollers to prevent the door from opening unexpectedly.

Step 2: Release the Tension on the Springs
1. Locate the winding cone on the end of the spring you're replacing. This is the part of the spring that connects to the winding bracket.
2. Insert a winding bar into one of the holes in the winding cone. Hold the bar firmly with both hands.
3. Using a wrench or socket, loosen the set screws on the winding cone until the spring begins to unwind. Keep a firm grip on the winding bar, as it will rotate as the spring releases tension.
4. Slowly allow the winding bar to rotate until all tension is released from the spring. You may need to use your body weight to control the unwinding process.
5. Remove the winding bar and repeat the process on the other end of the spring, if necessary.

Step 3: Remove the Old Spring
1. Once the tension is released, remove the bolts or nuts that secure the stationary cone to the center bracket.
2. Carefully remove the spring from the torsion shaft. If you have multiple springs, be sure to remove only one at a time to avoid confusion during reassembly.
3. If your springs are attached to a center bearing plate, remove the bolts securing the plate to the center bracket and slide the bearing off the torsion shaft.

Step 4: Install the New Spring
1. Slide the new spring onto the torsion shaft, ensuring that the stationary cone is facing the center bracket.
2. If you're replacing a center bearing, slide the new bearing onto the torsion shaft and secure it to the center bracket with the appropriate bolts.

3. Secure the stationary cone to the center bracket using the bolts or nuts removed earlier.

4. Repeat the process for the other spring, if applicable.

Step 5: Wind the New Spring

1. Insert a winding bar into the winding cone of the new spring.

2. Rotate the winding bar to apply tension to the spring. The number of turns required will depend on the specific spring you're installing. Refer to the manufacturer's instructions or consult with a professional for guidance.

3. As you wind the spring, keep a firm grip on the winding bar and maintain control of the winding process. Ensure that the spring is winding in the proper direction (upward) based on the end of the door you're working on.

4. Once the specified number of turns has been applied, tighten the set screws on the winding cone to lock the spring in place.

5. Repeat the winding process for the other spring, if applicable.

Step 6: Test and Adjust

1. Remove the C-clamp or locking pliers from the track, allowing the door to move freely.

2. Lift the garage door manually to check its balance. The door should remain in place when lifted to waist height and released. If it falls or rises on its own, you may need to adjust the spring tension.

3. To adjust the spring tension, repeat the winding or unwinding process in small increments (quarter turns) until the door is properly balanced.

4. Once the door is balanced, reconnect the garage door opener (if applicable) and test the door's operation.

Step 7: Clean Up and Maintain

1. Remove any tools or debris from the work area and dispose of the old springs properly.

2. Lubricate the torsion springs and other moving parts of the garage door with a silicone-based lubricant to ensure smooth operation and extend the life of your new springs.

3. Regularly inspect your garage door and springs for signs of wear or damage, and address any issues promptly to prevent more serious problems from developing.

Conclusion

Replacing torsion springs on your garage door is a complex and potentially hazardous task that requires careful planning, the right tools, and a strong emphasis on safety. By following the step-by-step guide provided above and prioritizing safety at every stage of the process, you can successfully replace your torsion springs and restore your garage door to optimal functionality.

However, it's important to recognize the inherent risks involved in working with torsion springs. If you're not entirely confident in your ability to safely complete the replacement process, or if you encounter any issues or complications along the way, don't hesitate to contact a professional garage door technician for assistance. They have the knowledge, experience, and tools necessary to replace your torsion springs safely and efficiently.

By properly maintaining your garage door and addressing any issues with your torsion springs promptly, you can ensure the longevity, reliability, and safety of your garage door system for years to come.

Step-by-Step Guide to Replacing Extension Springs

Extension springs are a common type of garage door spring, typically found on older garage door systems or those with limited overhead space. While replacing extension springs is generally less complex than replacing torsion springs, it still requires careful attention to safety and proper technique. In this section, we'll provide a detailed, step-by-step guide to replacing extension springs.

Safety Warning: Extension springs are under high tension and can cause serious injury if mishandled. Always wear safety goggles and gloves when working with extension springs, and never attempt to remove or adjust springs without first releasing the tension. If you're not confident in your ability to safely replace your extension springs, contact a professional garage door technician for assistance.

Tools and Materials Needed:
- New extension springs (properly sized for your door)
- Safety cables (if not already installed)
- Open-end wrench set
- Pliers
- Vise grips
- Ladder
- Safety goggles and gloves
- Tape measure
- Marker or pencil

Step 1: Prepare the Garage Door
1. Close the garage door and unplug the garage door opener (if applicable) to prevent accidental operation during the repair process.

2. Place a C-clamp or locking pliers on the track just above one of the rollers to prevent the door from opening unexpectedly.

Step 2: Release the Tension on the Springs
1. Locate the extension springs on either side of the garage door, running parallel to the horizontal tracks.
2. Attach a pair of vise grips or locking pliers to the track below the bottom roller on the side of the door with the spring you're replacing.
3. Using an open-end wrench, loosen the nut that secures the cable to the bracket at the bottom of the door.
4. Carefully remove the cable from the bracket, allowing the spring to extend fully and release its tension. Be prepared for the spring to expand rapidly as the tension is released.
5. Repeat the process on the other side of the door if you're replacing both springs.

Step 3: Remove the Old Spring
1. With the tension released, detach the safety cable (if present) from the spring and the track bracket.
2. Remove the spring from the track bracket and the rear mounting bracket.
3. If your springs are equipped with pulleys, remove the pulley and cable from the old spring.
4. Repeat the process for the other spring, if necessary.

Step 4: Install the New Spring
1. If your new springs did not come with pre-installed pulleys, attach the pulleys and cables to the new springs.
2. Attach one end of the new spring to the rear mounting bracket.
3. Stretch the spring toward the front of the garage door and connect it to the track bracket.
4. Repeat the process for the other spring, if applicable.

Step 5: Reconnect the Cables and Adjust Tension
1. Thread the cable through the pulley (if applicable) and reconnect it to the bracket at the bottom of the door.
2. Tighten the nut securing the cable to the bracket, ensuring that the cable is properly aligned and not twisted.
3. Remove the vise grips or locking pliers from the track, allowing the spring to bear the weight of the door.
4. Repeat the process on the other side of the door, if necessary.

Step 6: Install Safety Cables (if not already present)
1. Safety cables are essential for preventing injury in case of spring breakage. If your garage door doesn't already have safety cables, install them now.
2. Thread the safety cable through the center of the extension spring, securing one end to the rear mounting bracket and the other end to the track bracket.
3. Ensure that the safety cable is taut and properly positioned to contain the spring in case of breakage.
4. Repeat the process for the other spring, if applicable.

Step 7: Test and Adjust
1. Remove the C-clamp or locking pliers from the track, allowing the door to move freely.
2. Lift the garage door manually to check its balance. The door should remain in place when lifted to waist height and released. If it falls or rises on its own, you may need to adjust the spring tension.
3. To adjust the spring tension, move the cable to a different hole on the track bracket. Moving the cable to a higher hole will increase tension, while moving it to a lower hole will decrease tension.
4. Once the door is balanced, reconnect the garage door opener (if applicable) and test the door's operation.

Step 8: Clean Up and Maintain

1. Remove any tools or debris from the work area and dispose of the old springs properly.
2. Lubricate the extension springs and other moving parts of the garage door with a silicone-based lubricant to ensure smooth operation and extend the life of your new springs.
3. Regularly inspect your garage door and springs for signs of wear or damage, and address any issues promptly to prevent more serious problems from developing.

Conclusion

Replacing extension springs on your garage door requires careful planning, the right tools, and a commitment to safety. By following the step-by-step guide provided above and prioritizing safety throughout the process, you can successfully replace your extension springs and ensure that your garage door operates smoothly and reliably.

However, it's crucial to recognize the potential risks involved in working with extension springs. If you're not completely comfortable with the replacement process or encounter any difficulties along the way, don't hesitate to reach out to a professional garage door technician for assistance. They have the expertise and tools necessary to replace your extension springs safely and efficiently.

By properly maintaining your garage door and addressing any issues with your extension springs in a timely manner, you can ensure the longevity, reliability, and safety of your garage door system for years to come.

Adjusting Spring Tension for Optimal Performance

Proper spring tension is essential for the safe and efficient operation of your garage door. When the tension is correctly adjusted, your garage door should be balanced, meaning it should remain in place when manually lifted to waist height and released. If the door falls or rises on its own, or if it feels unusually heavy or light when operated manually, it's likely that the spring tension needs to be adjusted. In this section, we'll discuss how to adjust spring tension for both torsion and extension springs to achieve optimal performance.

Safety Warning: Adjusting garage door spring tension can be dangerous, as springs are under high tension and can cause serious injury if mishandled. If you're not confident in your ability to safely adjust your garage door springs, contact a professional garage door technician for assistance.

Adjusting Torsion Spring Tension
Tools Needed:
- Winding bars (two solid steel rods, typically 18" long)
- Adjustable wrench or socket set
- Safety goggles and gloves

Step 1: Prepare the Garage Door
1. Close the garage door and unplug the garage door opener (if applicable) to prevent accidental operation during the adjustment process.
2. Place a C-clamp or locking pliers on the track just above one of the rollers to prevent the door from opening unexpectedly.

Step 2: Locate the Adjustment Collar
1. Identify the adjustment collar, which is a ring with several holes located near the winding cone on the torsion spring.

2. The adjustment collar is used to increase or decrease spring tension by rotating the spring.

Step 3: Adjust the Spring Tension
1. Insert a winding bar into one of the holes in the adjustment collar.
2. While firmly holding the winding bar, loosen the set screws on the adjustment collar using an adjustable wrench or socket.
3. To increase tension, rotate the adjustment collar upward. To decrease tension, rotate the adjustment collar downward. Make adjustments in small increments (quarter turns) to avoid overloading the spring.
4. After each adjustment, retighten the set screws on the adjustment collar to lock the spring in place.
5. Repeat the process for the other torsion spring, if applicable, ensuring that both springs are adjusted equally.

Step 4: Test and Fine-tune
1. Remove the C-clamp or locking pliers from the track, allowing the door to move freely.
2. Manually lift the garage door to waist height and release it. The door should remain in place. If it falls or rises on its own, make further adjustments to the spring tension as needed.
3. Once the door is balanced, reconnect the garage door opener (if applicable) and test the door's operation.

Adjusting Extension Spring Tension
Tools Needed:
- Open-end wrench set
- Safety goggles and gloves

Step 1: Prepare the Garage Door
1. Close the garage door and unplug the garage door opener (if applicable) to prevent accidental operation during the adjustment process.

2. Place a C-clamp or locking pliers on the track just above one of the rollers to prevent the door from opening unexpectedly.

Step 2: Locate the Adjustment Bracket
1. Identify the adjustment bracket, which is a metal plate with several holes located at the front end of the extension spring assembly.
2. The adjustment bracket is used to increase or decrease spring tension by changing the position of the cable.

Step 3: Adjust the Spring Tension
1. Using an open-end wrench, loosen the nut that secures the cable to the adjustment bracket.
2. To increase tension, move the cable to a higher hole on the adjustment bracket. To decrease tension, move the cable to a lower hole.
3. Retighten the nut to secure the cable in its new position.
4. Repeat the process for the other extension spring, ensuring that both springs are adjusted equally.

Step 4: Test and Fine-tune
1. Remove the C-clamp or locking pliers from the track, allowing the door to move freely.
2. Manually lift the garage door to waist height and release it. The door should remain in place. If it falls or rises on its own, make further adjustments to the spring tension as needed.
3. Once the door is balanced, reconnect the garage door opener (if applicable) and test the door's operation.

Conclusion

Adjusting garage door spring tension is a crucial aspect of maintaining the safety, efficiency, and longevity of your garage door system. By following the steps outlined above and making adjustments in small increments, you can achieve optimal performance and ensure that your garage door operates smoothly and reliably.

However, it's essential to prioritize safety when adjusting spring tension. Garage door springs are under high tension and can cause serious injury if mishandled. If you're not entirely comfortable with the adjustment process or encounter any difficulties, don't hesitate to contact a professional garage door technician for assistance.

Regular maintenance, including periodic checks of spring tension and balance, can help prevent more serious issues from developing and extend the life of your garage door system. By staying proactive and addressing any issues promptly, you can ensure that your garage door continues to provide safe, reliable operation for years to come.

Chapter 2
Garage Door Openers

Types of Garage Door Openers

Garage door openers are motorized devices that automatically open and close garage doors, providing convenience and security for homeowners. They have become an essential part of modern homes, offering a range of features and benefits that make accessing your garage easier and safer. In this section, we'll explore the various types of garage door openers available, their key characteristics, and the advantages and disadvantages of each.

Types of Garage Door Openers

There are four main types of garage door openers, each with its own unique features and benefits:

1. Chain-Driven Openers
Chain-driven garage door openers are the most common and economical option. They use a metal chain to push or pull a trolley connected to the garage door, thus lifting and lowering the door.

Advantages:
- Affordable: Chain-driven openers are typically the least expensive option, making them a popular choice for budget-conscious homeowners.
- Reliable: These openers are known for their durability and reliability, as the metal chain is strong and can withstand heavy use.
- Suitable for heavy doors: Chain-driven openers are capable of lifting heavy garage doors, making them a good choice for oversized or wooden doors.

Disadvantages:
- Noise: The metal chain can be noisy during operation, which may be a concern for homeowners with living spaces adjacent to the garage.
- Maintenance: The chain requires regular lubrication to prevent rust and ensure smooth operation.
- Aesthetics: Some homeowners may find the appearance of the exposed chain less visually appealing compared to other types of openers.

2. Belt-Driven Openers

Belt-driven garage door openers use a reinforced rubber belt instead of a metal chain to lift and lower the garage door. The belt is connected to a trolley that moves along a track, similar to chain-driven openers.

Advantages:
- Quiet operation: The rubber belt is significantly quieter than a metal chain, making belt-driven openers a good choice for homes with living spaces above or adjacent to the garage.
- Smooth operation: The belt provides a smoother, more vibration-free operation compared to chain-driven openers.
- Low maintenance: Rubber belts require less maintenance than metal chains, as they don't need regular lubrication.

Disadvantages:
- Higher cost: Belt-driven openers are typically more expensive than chain-driven openers due to the cost of the reinforced rubber belt.
- Reduced strength: While suitable for most residential garage doors, belt-driven openers may not be as strong as chain-driven openers for lifting exceptionally heavy doors.
- Temperature sensitivity: Extreme temperatures can cause the rubber belt to expand or contract, potentially affecting performance.

3. Screw-Driven Openers
Screw-driven garage door openers use a threaded steel rod to lift and lower the garage door. The opener's motor rotates the rod, which in turn moves a trolley connected to the door.

Advantages:
- Strength: Screw-driven openers are known for their lifting power, making them suitable for heavy garage doors.
- Speed: These openers typically lift and lower doors faster than chain-driven or belt-driven openers.
- Low maintenance: The threaded steel rod requires minimal maintenance, as it doesn't need regular lubrication like chains or belts.

Disadvantages:
- Noise: Screw-driven openers can be noisy due to the steel rod's rotation, although they are generally quieter than chain-driven openers.
- Higher cost: These openers are often more expensive than chain-driven and belt-driven options.
- Climate sensitivity: Screw-driven openers may be affected by extreme temperature changes, which can cause the steel rod to expand or contract and potentially lead to misalignment.

4. Direct-Drive Openers
Direct-drive garage door openers, also known as jackshaft openers, are mounted on the wall beside the garage door rather than on the ceiling. They use a motor to directly turn the torsion spring, which lifts and lowers the door.

Advantages:
- Space-saving: By mounting on the wall, direct-drive openers free up ceiling space, making them ideal for garages with limited overhead room or those with cathedral ceilings.

- Quiet operation: The absence of chains, belts, or screws results in exceptionally quiet operation.
- Aesthetics: With no visible moving parts, direct-drive openers offer a clean, streamlined appearance.

Disadvantages:
- Higher cost: Direct-drive openers are typically the most expensive option due to their specialized design and installation requirements.
- Installation complexity: These openers require professional installation, as they must be precisely aligned with the torsion spring and connected to the garage door's drum.
- Limited compatibility: Direct-drive openers are not compatible with all garage door types and may require additional modifications for proper installation.

Conclusion

When choosing a garage door opener, consider factors such as your budget, the weight and size of your garage door, noise levels, maintenance requirements, and the available space in your garage. Each type of opener has its own set of advantages and disadvantages, so it's essential to select the one that best meets your specific needs and preferences.

Regardless of the type of opener you choose, proper installation, regular maintenance, and adherence to safety guidelines are crucial for ensuring the longevity and reliable performance of your garage door system. If you're unsure about which type of opener is best suited for your home, or if you require assistance with installation or maintenance, don't hesitate to consult with a professional garage door technician.

Choosing the Right Opener for Your Needs

Selecting the right garage door opener is essential for ensuring the convenience, safety, and smooth operation of your garage door system. With various types of openers available, each with its own set of features and benefits, it's important to consider your specific needs and preferences when making a decision. In this section, we'll discuss the key factors to keep in mind when choosing the right garage door opener for your home.

1. Type of Garage Door
The first factor to consider when selecting a garage door opener is the type of garage door you have. Different openers are designed to handle specific door types, weights, and sizes.

- Single-panel doors: These doors are typically lighter and easier to lift, making them compatible with most types of openers.
- Sectional doors: Sectional doors are the most common type of residential garage door and are compatible with all types of openers. However, heavier sectional doors may require a more powerful opener, such as a chain-driven or screw-driven model.
- Custom or oversized doors: If you have a custom or oversized garage door, consult with a professional to determine the most suitable opener type and horsepower rating.

2. Horsepower Rating
The horsepower (HP) rating of a garage door opener indicates its lifting capacity. Choosing an opener with the appropriate horsepower is crucial for ensuring reliable operation and prolonging the life of your garage door system.

- 1/2 HP: Suitable for most standard single-car garage doors.
- 3/4 HP: Recommended for heavier single-car doors or standard double-car garage doors.
- 1 HP and above: Ideal for oversized, heavy, or custom garage doors.

3. Noise Level

If you have living spaces above or adjacent to your garage, noise levels may be a significant concern. Consider the noise output of different opener types when making your selection.

- Belt-driven openers: These openers are the quietest option, making them ideal for homes with living spaces near the garage.
- Direct-drive openers: Like belt-driven openers, direct-drive models operate quietly due to the absence of chains or screws.
- Screw-driven and chain-driven openers: These openers tend to be louder than belt-driven and direct-drive models. If noise is a concern, consider adding vibration isolation kits or choosing a quieter option.

4. Safety and Security Features

Modern garage door openers come with a range of safety and security features designed to protect your family and property. When selecting an opener, look for the following features:

- Safety sensors: Photoelectric sensors detect obstacles in the door's path and reverse the door's motion to prevent accidents.
- Rolling code technology: This feature changes the opener's access code each time it's used, preventing unauthorized access to your garage.
- Battery backup: A backup battery ensures that your opener will continue to function during power outages.

- Smartphone compatibility: Some openers allow you to control and monitor your garage door using a smartphone app, providing added convenience and security.

5. Installation and Maintenance Requirements

Consider the installation and maintenance requirements of different opener types when making your decision.

- Chain-driven and belt-driven openers: These openers are relatively easy to install and maintain, although chain-driven models require regular lubrication.
- Screw-driven openers: While screw-driven openers require minimal maintenance, they may be more challenging to install due to their weight and size.
- Direct-drive openers: These openers require professional installation and may have limited compatibility with certain garage door types.

6. Budget

Garage door openers vary in price based on their type, features, and horsepower rating. Establish a budget and look for an opener that offers the best combination of features and performance within your price range.

- Chain-driven openers: Generally the most affordable option.
- Belt-driven and screw-driven openers: These openers are typically mid-range in terms of price.
- Direct-drive openers: Often the most expensive option due to their specialized design and installation requirements.

Conclusion

Choosing the right garage door opener involves careful consideration of your garage door type, horsepower requirements, noise level preferences, desired safety and security features, installation and maintenance needs, and budget. By weighing these factors and consulting with a professional garage door technician if needed, you can select an opener that will provide reliable, safe, and convenient operation for years to come.

Remember, proper installation and regular maintenance are essential for ensuring the longevity and optimal performance of your chosen garage door opener. By following the manufacturer's guidelines and addressing any issues promptly, you can maximize the benefits of your investment and enjoy a smoothly functioning garage door system.

Installing a New Garage Door Opener

Installing a new garage door opener can seem like a daunting task, but with the right tools, instructions, and safety precautions, it can be a manageable DIY project for many homeowners. However, if you're not confident in your ability to complete the installation safely, it's always best to hire a professional garage door technician. In this section, we'll provide a step-by-step guide to installing a new garage door opener, focusing on belt-driven and chain-driven models, as they are the most common types.

Safety Warning: Before beginning the installation process, ensure that you have disconnected the power to the garage door opener and the garage door itself to prevent any accidental operation or injury.

Tools and Materials Needed:
- New garage door opener (belt-driven or chain-driven)
- Ladder
- Adjustable wrench
- Socket wrench set
- Drill and drill bits
- Hammer
- Screwdrivers (flathead and Phillips)
- Pliers
- Level
- Tape measure
- Pencil
- Safety glasses and gloves

Step 1: Assemble the Opener and Rail
1. Unpack the new garage door opener and locate the installation manual. Familiarize yourself with the parts and instructions specific to your model.

2. Assemble the rail according to the manufacturer's instructions. This typically involves connecting the rail sections, attaching the trolley, and securing the belt or chain.
3. Install the idler pulley and tension the belt or chain as specified in the manual.

Step 2: Install the Header Bracket
1. Locate the center of the garage door and mark it on the header wall above the door.
2. Measure and mark a point 2-3 inches above the highest point of the garage door's travel (consult your specific model's instructions for the exact measurement).
3. Center the header bracket on the marked point and attach it securely to the header wall using lag bolts.

Step 3: Attach the Rail to the Header Bracket
1. Raise the assembled opener and rail, and position the rail onto the header bracket.
2. Align the rail with the center of the garage door and secure it to the header bracket using the clevis pin and cotter pin provided.

Step 4: Mount the Opener Motor
1. Raise the opener motor and align it with the mounting holes on the ceiling or wall (depending on your specific model).
2. Secure the opener motor to the mounting surface using the provided hardware. Ensure that the motor is level and properly aligned with the rail.

Step 5: Install the Safety Sensors
1. Attach the safety sensor brackets to the bottom of each side of the garage door frame, approximately 6 inches above the floor.
2. Secure the safety sensors to the brackets and align them, so they face each other across the garage door opening.
3. Run the wires from the sensors to the opener motor, following the manufacturer's wiring diagram.

Step 6: Connect the Door Arm

1. Attach the curved door arm to the trolley using the provided clevis pin and cotter pin.
2. Connect the straight door arm to the garage door bracket (usually pre-installed on the door).
3. Join the curved and straight door arms together, adjusting their length as needed to ensure a proper connection.

Step 7: Wire the Opener

1. Connect the safety sensor wires to the appropriate terminals on the opener motor, following the wiring diagram.
2. Plug the opener motor into a nearby outlet or hardwire it to an electrical box (if required by local building codes).

Step 8: Program the Opener and Remote Controls

1. Set the limit switches on the opener motor according to the manufacturer's instructions. This ensures that the door stops at the correct points when fully open and closed.
2. Program the remote controls and any keyless entry pads following the instructions provided with your specific model.

Step 9: Test and Adjust

1. Test the garage door opener by running it through a complete open and close cycle. Ensure that the door moves smoothly and stops at the correct positions.
2. Test the safety sensors by placing an object in the door's path while it's closing. The door should reverse when the object breaks the sensor beam.
3. Make any necessary adjustments to the limit switches, safety sensors, or door arm to ensure proper operation.

Conclusion

Installing a new garage door opener requires careful planning, attention to detail, and a commitment to following the manufacturer's instructions and safety guidelines. By following the step-by-step guide provided and ensuring proper installation of all components, you can enjoy the convenience and security of a properly functioning garage door opener.

Remember, if at any point during the installation process you feel unsure or uncomfortable, don't hesitate to contact a professional garage door technician for assistance. They can help ensure that your new opener is installed safely and correctly, providing you with peace of mind and reliable operation for years to come.

Once your new garage door opener is installed, be sure to perform regular maintenance, such as lubricating moving parts and checking for wear or damage, to keep your garage door system running smoothly and extend its lifespan.

Programming and Adjusting Your Opener

After installing your new garage door opener, it's essential to properly program and adjust it to ensure safe, reliable, and convenient operation. This process involves setting the travel limits, programming the remote controls and keyless entry pads, and adjusting the force and sensitivity settings. In this section, we'll provide a detailed guide on programming and adjusting your garage door opener.

Safety Warning: Before making any adjustments to your garage door opener, ensure that you have disconnected the power to the opener to prevent any accidental operation or injury.

Tools and Materials Needed:
- Garage door opener manual
- Remote controls
- Keyless entry pad (if applicable)
- Flathead and Phillips screwdrivers
- Stepladder

Step 1: Set the Travel Limits
Travel limits determine where the garage door stops when it is fully open and fully closed. Proper adjustment of these limits is crucial for ensuring that the door operates safely and does not cause damage to the opener or the door itself.

1. Locate the limit adjustment screws or knobs on the opener motor. These are typically labeled "up" and "down" or "open" and "close."
2. Using a screwdriver, adjust the "down" or "close" limit screw until the door closes fully without excessive force or gaps between the door and the floor.

3. Adjust the "up" or "open" limit screw until the door opens fully without the trolley hitting the motor unit or the door extending beyond the opening.
4. Test the door's operation by running it through a complete open and close cycle, making further adjustments as needed.

Step 2: Program Remote Controls
Programming your remote controls allows you to conveniently open and close your garage door from inside your vehicle.

1. Locate the "learn" button on the opener motor. This button is typically located near the wiring terminals and may be labeled "learn," "program," or with a specific color.
2. Press and release the "learn" button. An LED light near the button should light up or blink, indicating that the opener is in learning mode.
3. Within 30 seconds, press and hold the desired button on the remote control until the opener light blinks or the motor clicks, signaling that the code has been accepted.
4. Repeat the process for any additional remote controls or buttons you wish to program.

Step 3: Program Keyless Entry Pad (if applicable)
A keyless entry pad allows you to open and close your garage door by entering a personalized code, providing an alternative to using remote controls.

1. Locate the "learn" button on the opener motor, as described in Step 2.
2. Press and release the "learn" button. The LED light should light up or blink.
3. Within 30 seconds, enter a 4-digit personal identification number (PIN) of your choice on the keyless entry pad, then press and hold the "enter" or "star" key until the opener light blinks or the motor clicks.

4. Test the keyless entry pad by entering your PIN and verifying that the door opens or closes accordingly.

Step 4: Adjust Force and Sensitivity Settings

Force and sensitivity settings control the amount of power the opener motor uses to lift and lower the garage door. Proper adjustment of these settings helps ensure safe operation and prevents damage to the door or opener.

1. Locate the force adjustment screws or knobs on the opener motor. These are typically labeled "force," "sensitivity," or with specific colors.
2. Using a screwdriver, slightly adjust the force settings according to the manufacturer's instructions. Typically, turning the screws clockwise increases the force, while turning them counterclockwise decreases it.
3. Test the door's operation by placing a 2x4 board flat on the floor in the door's path. The door should reverse when it contacts the board. If it doesn't, reduce the force setting and test again.
4. Repeat the process until the door reverses properly upon contact with the board, without applying excessive force.

Step 5: Test Safety Features

Ensuring that your garage door opener's safety features are functioning properly is essential for preventing accidents and injuries.

1. Test the safety sensors by placing an object in the door's path while it's closing. The door should reverse when the object breaks the sensor beam. If it doesn't, check the sensor alignment and wiring, and make adjustments as needed.

2. Test the manual release handle by pulling it towards you. This should disengage the trolley from the opener, allowing you to manually open and close the door. Ensure that the trolley reconnects properly when the door is opened again using the opener.

Conclusion

Programming and adjusting your garage door opener is a critical step in ensuring its safe, reliable, and convenient operation. By setting the travel limits, programming the remote controls and keyless entry pad, adjusting the force and sensitivity settings, and testing the safety features, you can optimize your opener's performance and minimize the risk of accidents or damage.

Remember to consult your opener's manual for specific instructions and safety guidelines, as procedures may vary slightly between models and brands. If you encounter any issues or are unsure about making adjustments, don't hesitate to contact a professional garage door technician for assistance.

Regular maintenance, such as lubricating moving parts, checking for wear or damage, and testing safety features, can help prolong the life of your garage door opener and ensure its continued safe operation. By staying proactive and addressing any issues promptly, you can enjoy the convenience and security of a properly functioning garage door opener for years to come.

Troubleshooting Common Opener Issues

Even with proper installation and maintenance, garage door openers can sometimes experience issues that disrupt their normal operation. Knowing how to troubleshoot common opener problems can help you quickly identify and resolve issues, saving you time and money. In this section, we'll discuss some of the most common garage door opener issues and provide step-by-step guidance on how to diagnose and fix them.

Safety Warning: Before attempting any troubleshooting or repairs, always ensure that you have disconnected the power to the garage door opener to prevent any accidental operation or injury.

Tools and Materials Needed:
- Garage door opener manual
- Screwdrivers (flathead and Phillips)
- Pliers
- Multimeter
- Ladder
- Lubricant (silicone-based or as recommended by the manufacturer)

Problem 1: Garage Door Doesn't Open or Close
If your garage door fails to open or close when prompted, there could be several underlying causes.

Troubleshooting Steps:
1. Check the power supply: Ensure that the opener is plugged in and that the outlet is functioning properly. If necessary, test the outlet with another device.
2. Check the remote control batteries: Replace the batteries in your remote control and test the door's operation.

3. Check the safety sensors: Ensure that the safety sensors are properly aligned and free from obstructions. Clean the sensor lenses with a soft cloth if needed.

4. Check for manual lock engagement: Verify that the manual lock on the garage door is not engaged, preventing the opener from operating.

5. Check the door's balance: Disconnect the opener from the door and manually lift the door halfway. If the door doesn't remain in place, it may be unbalanced, causing the opener to strain or fail to operate properly. Adjust the door's balance or contact a professional for assistance.

Problem 2: Garage Door Reverses Unexpectedly
If your garage door reverses direction unexpectedly while closing, it may be due to safety sensor issues or incorrect force settings.

Troubleshooting Steps:
1. Check the safety sensors: Ensure that the safety sensors are properly aligned, clean, and free from obstructions. If the sensors are misaligned or obstructed, the door will reverse to prevent accidents.

2. Adjust the force settings: If the force settings are set too low, the door may interpret normal resistance as an obstruction and reverse. Adjust the force settings according to the manufacturer's instructions, as outlined in the "Programming and Adjusting Your Opener" section.

3. Check for physical obstructions: Inspect the garage door's path for any objects or debris that could be causing the door to reverse. Remove any obstructions found.

Problem 3: Garage Door Opens or Closes Partially
If your garage door only opens or closes partially, it may be due to incorrect limit settings or obstruction issues.

Troubleshooting Steps:
1. Adjust the travel limits: Check and adjust the travel limit settings as described in the "Programming and Adjusting Your Opener" section. Ensure that the limits are set correctly for your door's full open and close positions.
2. Check for physical obstructions: Inspect the garage door's track, rollers, and hinges for any obstructions, damage, or binding. Remove obstructions and repair or replace damaged components as needed.
3. Lubricate moving parts: Apply a silicone-based lubricant to the door's rollers, hinges, and track to ensure smooth operation and minimize resistance.

Problem 4: Garage Door Opener Makes Unusual Noises
Unusual noises coming from your garage door opener can indicate various issues, such as worn-out parts, loose hardware, or inadequate lubrication.

Troubleshooting Steps:
1. Identify the type of noise: Listen carefully to the noise and note its characteristics (e.g., grinding, squeaking, or rattling) to help determine the potential cause.
2. Tighten loose hardware: Inspect the opener's mounting hardware, track, and door hinges for any loose bolts or screws. Tighten them securely to eliminate rattling noises.
3. Lubricate moving parts: Apply a silicone-based lubricant to the opener's chain or belt, as well as the door's rollers, hinges, and track. Proper lubrication can help reduce noise and ensure smooth operation.
4. Replace worn-out parts: If the noise persists after lubrication and tightening hardware, it may indicate worn-out gears, bearings, or sprockets in the opener. Contact a professional garage door technician to assess and replace the affected components.

Problem 5: Keyless Entry Pad or Remote Control Not Working

If your keyless entry pad or remote control fails to operate the garage door opener, it may be due to programming issues, dead batteries, or signal interference.

Troubleshooting Steps:
1. Replace batteries: For battery-operated devices, replace the batteries with fresh ones and test the device again.
2. Reprogram the device: If the keyless entry pad or remote control has lost its programming, follow the manufacturer's instructions to reprogram the device, as described in the "Programming and Adjusting Your Opener" section.
3. Check for signal interference: Ensure that there are no nearby devices (e.g., appliances or electronics) that could be causing signal interference. Move the affected device closer to the opener and test its operation.

Conclusion

Troubleshooting common garage door opener issues involves carefully assessing the problem, identifying potential causes, and following a systematic approach to resolve the issue. By referring to your opener's manual and the troubleshooting steps outlined in this section, you can often diagnose and fix minor problems on your own.

However, if the issue persists or you're unsure about any aspect of the troubleshooting process, don't hesitate to contact a professional garage door technician for assistance. They have the expertise and tools necessary to identify and resolve more complex opener issues safely and effectively.

Regular maintenance, such as lubricating moving parts, checking for wear or damage, and testing safety features, can help prevent many common opener issues from occurring in the first place. By staying proactive and addressing any concerns promptly, you can ensure the continued safe, reliable, and convenient operation of your garage door opener.

Chapter 3
Garage Door Tracks
Understanding Garage Door Track Components

Garage door tracks are a crucial component of your garage door system, providing a path for the door to travel along as it opens and closes. Understanding the various parts of your garage door tracks and their functions is essential for ensuring proper maintenance, troubleshooting, and repairs. In this section, we'll take a closer look at the key components that make up your garage door track system.

Understanding Garage Door Track Components

1. Vertical Tracks
Vertical tracks are the sections of the track system that run perpendicular to the ground on either side of the garage door opening. They are typically made of galvanized steel and are mounted to the jambs of the garage door frame.

Key Features:
- Guide the garage door as it moves up and down
- Support the door's weight when it is fully open
- Provide a mounting surface for the garage door's rollers
- Available in different lengths and gauges to accommodate various door sizes and weights

2. Horizontal Tracks
Horizontal tracks are the sections of the track system that run parallel to the ceiling, extending from the top of the vertical tracks towards the back of the garage. They are suspended from the ceiling using angle iron or steel supports.

Key Features:
- Guide the garage door as it moves along the ceiling when fully open
- Provide a stable platform for the door to rest on when open
- Ensure proper alignment and smooth operation of the door
- Available in different lengths and gauges to accommodate various door sizes and weights

3. Curved Track Sections

Curved track sections, also known as radius tracks, are the transitional pieces that connect the vertical tracks to the horizontal tracks. They are designed to provide a smooth, gradual transition for the garage door rollers as the door moves from the vertical to the horizontal position.

Key Features:
- Facilitate smooth and efficient operation of the garage door
- Ensure proper alignment of the door as it transitions between the vertical and horizontal tracks
- Available in various radii to accommodate different garage door configurations and space requirements

4. Mounting Hardware

Mounting hardware consists of the various brackets, hinges, and fasteners used to secure the garage door tracks to the door frame and ceiling.

Key Components:
- Jamb brackets: Used to attach the vertical tracks to the garage door frame
- Flag brackets: Mounted at the top of the vertical tracks to support the curved track sections and provide a mounting point for the garage door cables
- Track hangers: Used to suspend the horizontal tracks from the ceiling

- Fasteners: Include bolts, nuts, and screws used to secure the tracks and brackets in place

5. Rollers

Garage door rollers are the wheels that run along the tracks, allowing the door to move smoothly and efficiently. They are typically made of steel, nylon, or plastic and are mounted on the edges of the garage door panels.

Key Features:
- Facilitate smooth and quiet operation of the garage door
- Available in various sizes and materials to suit different door weights and track configurations
- Nylon and plastic rollers are quieter and more durable than steel rollers
- Some rollers feature ball bearings for enhanced performance and longevity

6. Track Reinforcement

In some cases, garage door tracks may require additional reinforcement to ensure stability and prevent bending or sagging. Reinforcement options include:

- Angle iron: Heavier gauge steel used to reinforce horizontal tracks and prevent sagging under the weight of the door
- Struts: Horizontal bars attached to the garage door panels to provide additional rigidity and prevent the door from bending or warping

Conclusion

A thorough understanding of your garage door track components is essential for proper maintenance, troubleshooting, and repairs. By familiarizing yourself with the vertical tracks, horizontal tracks, curved track sections, mounting hardware, rollers, and reinforcement options, you'll be better equipped to identify and address any issues that may arise with your garage door system.

Regular inspections of your garage door tracks can help you catch potential problems early, such as bent or misaligned tracks, worn-out rollers, or loose mounting hardware. By promptly addressing these issues, you can prevent more serious damage and ensure the safe, reliable, and efficient operation of your garage door.

If you encounter any problems with your garage door tracks or are unsure about performing maintenance or repairs, don't hesitate to contact a professional garage door technician. They have the expertise, tools, and experience necessary to diagnose and resolve track-related issues safely and effectively, helping to prolong the life of your garage door system.

Identifying Misaligned or Damaged Tracks

Properly aligned and adjusted horizontal garage door tracks are essential for ensuring the smooth, safe, and efficient operation of your garage door. Misaligned or poorly adjusted horizontal tracks can cause a range of issues, including uneven door movement, increased wear on components, and even complete door failure. In this section, we'll provide a detailed guide on how to align and adjust your garage door's horizontal tracks.

Safety Warning: Before attempting any adjustments to your garage door tracks, ensure that the door is fully closed and disconnected from the opener. Additionally, wear safety gloves and glasses to protect yourself from potential injuries.

Tools and Materials Needed:
- Measuring tape
- Level
- Socket wrench set
- Adjustable wrench
- Pliers
- Screwdriver
- Stepladder
- Lubricant (silicone-based or as recommended by the manufacturer)

Step 1: Inspect the Horizontal Tracks
Begin by visually inspecting the horizontal tracks for any signs of damage, such as bent or dented sections, rust, or loose mounting hardware.

1. If you notice any significant damage, such as severely bent tracks, it's best to contact a professional garage door technician for repairs or replacement.

2. Minor dents or bends can often be corrected using a rubber mallet or a track bending tool. Gently tap the affected area until it is straightened.

Step 2: Check Track Alignment

Proper alignment of the horizontal tracks is crucial for the garage door's smooth operation. To check the alignment, follow these steps:

1. Close the garage door and disconnect it from the opener.
2. Measure the distance between the left and right horizontal tracks at the front, middle, and rear of the garage door opening. The measurements should be equal at all three points.
3. If the measurements are not equal, the tracks are misaligned and require adjustment.

Step 3: Adjust Track Alignment

To adjust the alignment of the horizontal tracks, follow these steps:

1. Locate the mounting bolts and nuts that secure the horizontal track hangers to the ceiling.
2. Loosen the bolts slightly, allowing the track to be adjusted.
3. Gently tap the track with a rubber mallet to move it into the proper alignment. Use the measurements from Step 2 as a guide.
4. Once the track is properly aligned, retighten the mounting bolts securely.
5. Repeat the process for the other horizontal track, ensuring that both tracks are parallel and level.

Step 4: Check Track Spacing

In addition to alignment, it's important to ensure that the horizontal tracks are spaced correctly to allow the garage door to move smoothly.

1. Measure the distance between the inside edges of the horizontal tracks. This distance should be the same as the width of your garage door plus 1/4 inch on each side for clearance.
2. If the spacing is incorrect, loosen the mounting bolts and adjust the tracks as needed.
3. Retighten the bolts once the proper spacing is achieved.

Step 5: Lubricate the Tracks and Rollers
After aligning and adjusting the horizontal tracks, it's essential to lubricate the tracks and rollers to ensure smooth, quiet operation and prevent premature wear.

1. Apply a silicone-based lubricant or a lubricant recommended by the manufacturer to the inside surface of the horizontal tracks.
2. Lubricate the rollers by applying a few drops of lubricant to the roller axles or bearings.
3. Open and close the garage door manually a few times to distribute the lubricant evenly.

Step 6: Test the Door Operation
Once you have completed the alignment, adjustment, and lubrication process, it's time to test the garage door's operation.

1. Reconnect the garage door to the opener.
2. Open and close the door using the opener, observing its movement along the horizontal tracks.
3. The door should move smoothly and evenly, without binding or rubbing against the tracks.
4. If you notice any issues, such as uneven movement or rubbing, recheck the track alignment and spacing, and make further adjustments as needed.

Conclusion

Aligning and adjusting your garage door's horizontal tracks is a crucial maintenance task that helps ensure the safe, smooth, and reliable operation of your door. By regularly inspecting your tracks for damage, checking and adjusting their alignment and spacing, and lubricating the tracks and rollers, you can prevent a range of issues and extend the life of your garage door system.

Remember, if you encounter any significant damage or are unsure about performing any of the steps outlined in this guide, don't hesitate to contact a professional garage door technician for assistance. They have the knowledge, skills, and tools necessary to handle more complex track repairs and adjustments safely and effectively.

By staying proactive and addressing any track-related issues promptly, you can enjoy the convenience and security of a properly functioning garage door for years to come.

Aligning and Adjusting Horizontal Tracks

Properly aligned and adjusted horizontal garage door tracks are essential for ensuring the smooth, safe, and efficient operation of your garage door. Misaligned or poorly adjusted horizontal tracks can cause a range of issues, including uneven door movement, increased wear on components, and even complete door failure. In this section, we'll provide a detailed guide on how to align and adjust your garage door's horizontal tracks.

Safety Warning: Before attempting any adjustments to your garage door tracks, ensure that the door is fully closed and disconnected from the opener. Additionally, wear safety gloves and glasses to protect yourself from potential injuries.

Tools and Materials Needed:
- Measuring tape
- Level
- Socket wrench set
- Adjustable wrench
- Pliers
- Screwdriver
- Stepladder
- Lubricant (silicone-based or as recommended by the manufacturer)

Step 1: Inspect the Horizontal Tracks
Begin by visually inspecting the horizontal tracks for any signs of damage, such as bent or dented sections, rust, or loose mounting hardware.

1. If you notice any significant damage, such as severely bent tracks, it's best to contact a professional garage door technician for repairs or replacement.

2. Minor dents or bends can often be corrected using a rubber mallet or a track bending tool. Gently tap the affected area until it is straightened.

Step 2: Check Track Alignment
Proper alignment of the horizontal tracks is crucial for the garage door's smooth operation. To check the alignment, follow these steps:

1. Close the garage door and disconnect it from the opener.
2. Measure the distance between the left and right horizontal tracks at the front, middle, and rear of the garage door opening. The measurements should be equal at all three points.
3. If the measurements are not equal, the tracks are misaligned and require adjustment.

Step 3: Adjust Track Alignment
To adjust the alignment of the horizontal tracks, follow these steps:

1. Locate the mounting bolts and nuts that secure the horizontal track hangers to the ceiling.
2. Loosen the bolts slightly, allowing the track to be adjusted.
3. Gently tap the track with a rubber mallet to move it into the proper alignment. Use the measurements from Step 2 as a guide.
4. Once the track is properly aligned, retighten the mounting bolts securely.
5. Repeat the process for the other horizontal track, ensuring that both tracks are parallel and level.

Step 4: Check Track Spacing
In addition to alignment, it's important to ensure that the horizontal tracks are spaced correctly to allow the garage door to move smoothly.

1. Measure the distance between the inside edges of the horizontal tracks. This distance should be the same as the width of your garage door plus 1/4 inch on each side for clearance.
2. If the spacing is incorrect, loosen the mounting bolts and adjust the tracks as needed.
3. Retighten the bolts once the proper spacing is achieved.

Step 5: Lubricate the Tracks and Rollers
After aligning and adjusting the horizontal tracks, it's essential to lubricate the tracks and rollers to ensure smooth, quiet operation and prevent premature wear.

1. Apply a silicone-based lubricant or a lubricant recommended by the manufacturer to the inside surface of the horizontal tracks.
2. Lubricate the rollers by applying a few drops of lubricant to the roller axles or bearings.
3. Open and close the garage door manually a few times to distribute the lubricant evenly.

Step 6: Test the Door Operation
Once you have completed the alignment, adjustment, and lubrication process, it's time to test the garage door's operation.

1. Reconnect the garage door to the opener.
2. Open and close the door using the opener, observing its movement along the horizontal tracks.
3. The door should move smoothly and evenly, without binding or rubbing against the tracks.
4. If you notice any issues, such as uneven movement or rubbing, recheck the track alignment and spacing, and make further adjustments as needed.

Conclusion

Aligning and adjusting your garage door's horizontal tracks is a crucial maintenance task that helps ensure the safe, smooth, and reliable operation of your door. By regularly inspecting your tracks for damage, checking and adjusting their alignment and spacing, and lubricating the tracks and rollers, you can prevent a range of issues and extend the life of your garage door system.

Remember, if you encounter any significant damage or are unsure about performing any of the steps outlined in this guide, don't hesitate to contact a professional garage door technician for assistance. They have the knowledge, skills, and tools necessary to handle more complex track repairs and adjustments safely and effectively.

By staying proactive and addressing any track-related issues promptly, you can enjoy the convenience and security of a properly functioning garage door for years to come.

Aligning and Adjusting Vertical Tracks

Just like horizontal tracks, properly aligned and adjusted vertical garage door tracks are essential for the smooth, safe, and efficient operation of your garage door. Misaligned or poorly adjusted vertical tracks can lead to various issues, such as uneven door movement, increased wear on components, and even door jamming or derailment. In this section, we'll provide a comprehensive guide on how to align and adjust your garage door's vertical tracks.

Safety Warning: Before attempting any adjustments to your garage door tracks, ensure that the door is fully closed and disconnected from the opener. Additionally, wear safety gloves and glasses to protect yourself from potential injuries.

Tools and Materials Needed:
- Measuring tape
- Level
- Socket wrench set
- Adjustable wrench
- Pliers
- Screwdriver
- Stepladder
- Lubricant (silicone-based or as recommended by the manufacturer)

Step 1: Inspect the Vertical Tracks
Begin by visually inspecting the vertical tracks for any signs of damage, such as bent or dented sections, rust, or loose mounting hardware.

1. If you notice any significant damage, such as severely bent tracks, it's best to contact a professional garage door technician for repairs or replacement.

2. Minor dents or bends can often be corrected using a rubber mallet or a track bending tool. Gently tap the affected area until it is straightened.

Step 2: Check Track Alignment
Proper alignment of the vertical tracks is crucial for the garage door's smooth operation. To check the alignment, follow these steps:

1. Close the garage door and disconnect it from the opener.
2. Place a level vertically against the left and right vertical tracks, ensuring that the tracks are plumb (perfectly vertical).
3. If the tracks are not plumb, they require adjustment.

Step 3: Adjust Track Alignment
To adjust the alignment of the vertical tracks, follow these steps:

1. Locate the mounting brackets that secure the vertical tracks to the garage door frame.
2. Loosen the bolts or screws that hold the brackets in place, allowing the tracks to be adjusted.
3. Gently tap the tracks with a rubber mallet to move them into the proper alignment. Use the level as a guide to ensure the tracks are plumb.
4. Once the tracks are properly aligned, retighten the mounting bolts or screws securely.
5. Repeat the process for both vertical tracks, ensuring they are parallel to each other and plumb.

Step 4: Check Track Spacing
In addition to alignment, it's important to ensure that the vertical tracks are spaced correctly to allow the garage door to move smoothly.

1. Measure the distance between the inside edges of the vertical tracks at the top and bottom. This distance should be equal to the width of your garage door plus 1/4 inch on each side for clearance.
2. If the spacing is incorrect, loosen the mounting brackets and adjust the tracks as needed.
3. Retighten the brackets once the proper spacing is achieved.

Step 5: Inspect and Adjust the Weatherstripping

The weatherstripping along the vertical tracks helps seal the garage door and prevent drafts and pests from entering the garage.

1. Inspect the weatherstripping for any signs of wear, damage, or gaps.
2. If the weatherstripping is worn or damaged, replace it with new material.
3. Adjust the weatherstripping so that it fits snugly against the garage door when closed, creating an effective seal.

Step 6: Lubricate the Tracks and Rollers

After aligning and adjusting the vertical tracks, it's essential to lubricate the tracks and rollers to ensure smooth, quiet operation and prevent premature wear.

1. Apply a silicone-based lubricant or a lubricant recommended by the manufacturer to the inside surface of the vertical tracks.
2. Lubricate the rollers by applying a few drops of lubricant to the roller axles or bearings.
3. Open and close the garage door manually a few times to distribute the lubricant evenly.

Step 7: Test the Door Operation

Once you have completed the alignment, adjustment, and lubrication process, it's time to test the garage door's operation.

1. Reconnect the garage door to the opener.
2. Open and close the door using the opener, observing its movement along the vertical tracks.
3. The door should move smoothly and evenly, without binding or rubbing against the tracks.
4. If you notice any issues, such as uneven movement or rubbing, recheck the track alignment and spacing, and make further adjustments as needed.

Conclusion

Aligning and adjusting your garage door's vertical tracks is a critical maintenance task that helps ensure the safe, smooth, and reliable operation of your door. By regularly inspecting your tracks for damage, checking and adjusting their alignment and spacing, lubricating the tracks and rollers, and maintaining the weatherstripping, you can prevent a range of issues and extend the life of your garage door system.

Remember, if you encounter any significant damage or are unsure about performing any of the steps outlined in this guide, don't hesitate to contact a professional garage door technician for assistance. They have the knowledge, skills, and tools necessary to handle more complex track repairs and adjustments safely and effectively.

By staying proactive and addressing any track-related issues promptly, you can enjoy the convenience and security of a properly functioning garage door for years to come.

Lubricating and Maintaining Tracks for Smooth Operation

Regular lubrication and maintenance of your garage door tracks are essential for ensuring smooth, quiet, and efficient operation of your garage door. Properly lubricated tracks minimize friction, reduce wear on components, and help prevent rust and corrosion. In this section, we'll provide a detailed guide on how to lubricate and maintain your garage door tracks for optimal performance.

Safety Warning: Before lubricating or performing any maintenance on your garage door tracks, ensure that the door is fully closed and disconnected from the opener. Additionally, wear safety gloves and glasses to protect yourself from potential injuries.

Tools and Materials Needed:
- Clean, lint-free cloth
- Silicone-based lubricant or a lubricant recommended by the manufacturer
- Stiff-bristled brush
- Vacuum cleaner or compressed air
- Safety gloves and glasses

Step 1: Clean the Tracks
Before applying lubricant, it's essential to clean the tracks to remove any dirt, debris, or old lubricant buildup.

1. Close the garage door and disconnect it from the opener.
2. Using a stiff-bristled brush, scrub the inside of the tracks to loosen any dirt or debris.
3. Vacuum the tracks or use compressed air to remove the loosened debris.
4. Wipe the tracks clean with a lint-free cloth to remove any remaining dirt or residue.

Step 2: Inspect the Tracks

While cleaning the tracks, take the opportunity to inspect them for any signs of damage or wear.

1. Look for bent, dented, or misaligned sections of the track.
2. Check for rust or corrosion, particularly in areas exposed to moisture.
3. Ensure that all mounting hardware is tight and secure.
4. If you notice any significant damage or wear, contact a professional garage door technician for repairs or replacement.

Step 3: Lubricate the Tracks

Once the tracks are clean and inspected, it's time to apply lubricant.

1. Choose a silicone-based lubricant or a lubricant specifically recommended by the garage door manufacturer. Avoid using WD-40 or other all-purpose lubricants, as they can attract dust and dirt.
2. Apply a thin layer of lubricant to the inside surface of the tracks, using a clean, lint-free cloth or a spray applicator.
3. Focus on the areas where the rollers make contact with the tracks, as these are the most critical points for reducing friction.
4. Wipe away any excess lubricant to prevent dirt and debris from adhering to the tracks.

Step 4: Lubricate the Rollers and Hinges

In addition to lubricating the tracks, it's important to lubricate the garage door rollers and hinges to ensure smooth operation.

1. Apply a few drops of silicone-based lubricant to the roller axles or bearings.
2. Lubricate the hinges by applying a small amount of lubricant to the pivot points.

3. Open and close the garage door manually a few times to distribute the lubricant evenly.

Step 5: Maintain a Regular Lubrication Schedule
To keep your garage door tracks in optimal condition, establish a regular lubrication and maintenance schedule.

1. Lubricate the tracks, rollers, and hinges every six months or more frequently if you live in a humid or coastal environment.
2. Clean the tracks and inspect them for damage or wear at least once a year, or whenever you notice any issues with the door's operation.
3. Keep a record of your maintenance tasks to help you stay on track and identify any recurring issues.

Conclusion

Lubricating and maintaining your garage door tracks is a simple but essential task that helps ensure the smooth, quiet, and reliable operation of your door. By regularly cleaning and inspecting your tracks, applying the appropriate lubricant, and maintaining a consistent maintenance schedule, you can prevent a range of issues and extend the life of your garage door system.

Remember, if you encounter any significant damage or are unsure about performing any of the steps outlined in this guide, don't hesitate to contact a professional garage door technician for assistance. They have the knowledge, skills, and tools necessary to handle more complex track repairs and maintenance tasks safely and effectively.

By staying proactive and addressing any track-related issues promptly, you can enjoy the convenience and security of a properly functioning garage door for years to come.

Chapter 4
Garage Door Panels
Assessing Panel Damage

Garage door panels are the individual sections that make up the main body of your garage door. These panels are typically made from materials such as steel, aluminum, wood, or fiberglass, and they play a crucial role in the appearance, security, and insulation of your garage. Understanding how to assess panel damage and address any issues promptly is essential for maintaining the overall health and functionality of your garage door. In this section, we'll focus on assessing panel damage and provide guidance on identifying and addressing common panel issues.

Assessing Panel Damage

Regular inspections of your garage door panels can help you identify damage early, preventing minor issues from escalating into more serious and costly problems. Here's a step-by-step guide to assessing panel damage:

Step 1: Visual Inspection
Begin by conducting a thorough visual inspection of your garage door panels, both from the inside and outside of your garage.

1. Look for any obvious signs of damage, such as dents, cracks, rust, or rot, depending on the material of your panels.
2. Check for any gaps or misalignment between panels, as this can indicate issues with the hinges or tracks.
3. Inspect the paint or finish for any peeling, flaking, or fading, which can compromise the panel's protection against the elements.

Step 2: Physical Inspection

After completing the visual inspection, proceed with a physical inspection to assess the extent of any damage and identify potential underlying issues.

1. Gently press against each panel to check for any softness, sponginess, or give, which can indicate underlying structural damage or rot in wooden panels.
2. Test the stability of each panel by gently trying to shift it from side to side. Any movement or looseness can suggest issues with the hinges or mounting hardware.
3. Check the mounting hardware, such as hinges and brackets, for any signs of rust, corrosion, or looseness.

Step 3: Identifying the Cause of Damage

Once you've assessed the extent of the damage, try to identify the underlying cause to determine the best course of action for repair or replacement.

Common causes of panel damage include:
- Impact damage from vehicles or other objects
- Wear and tear from age and regular use
- Exposure to harsh weather conditions, such as extreme temperatures, humidity, or saltwater
- Improper maintenance, such as failing to clean or lubricate moving parts regularly
- Pest infestations, particularly in wooden panels

Step 4: Determining the Severity of Damage

Based on your assessment, determine the severity of the damage and whether repair or replacement is necessary.

- Minor damage, such as small dents or scrapes, can often be repaired with simple techniques like paintless dent repair or a fresh coat of paint.

- Moderate damage, such as larger dents or cracks, may require more extensive repairs, such as filling and sanding the damaged area before repainting.
- Severe damage, such as large holes, significant rust or rot, or structural instability, may necessitate replacing the affected panel(s) to maintain the integrity and safety of your garage door.

Step 5: Addressing Panel Damage
Once you've determined the severity of the damage and the necessary course of action, take steps to address the issue promptly.

- For minor repairs, such as filling small cracks or touching up paint, consult the manufacturer's guidelines and use appropriate materials and techniques.
- For more extensive repairs or panel replacements, it's often best to consult a professional garage door technician to ensure the work is done safely and correctly.
- If you decide to replace damaged panels yourself, be sure to follow the manufacturer's instructions carefully and prioritize safety throughout the process.

Conclusion

Regularly assessing your garage door panels for damage is an essential part of maintaining the overall health, appearance, and functionality of your garage door. By conducting thorough visual and physical inspections, identifying the cause and severity of any damage, and addressing issues promptly, you can prevent minor problems from escalating and extend the lifespan of your garage door.

Remember, if you encounter any damage that you're unsure how to address or that compromises the safety or structural integrity of your garage door, don't hesitate to contact a professional garage door technician for expert advice and assistance. By staying proactive and vigilant in your garage door maintenance, you can ensure that your door continues to provide reliable, secure, and efficient operation for years to come.

Repairing Minor Dents and Cracks

Minor dents and cracks in your garage door panels may seem like purely cosmetic issues, but they can lead to more serious problems if left unaddressed. These small imperfections can allow moisture to penetrate the panel, leading to rust, corrosion, or rot, depending on the material of your door. Additionally, dents and cracks can weaken the structural integrity of the panel over time, making it more susceptible to further damage. In this section, we'll provide a detailed guide on repairing minor dents and cracks in your garage door panels.

Safety Warning: Before attempting any repairs on your garage door panels, ensure that the door is fully closed and disconnected from the opener. Additionally, wear safety gloves and glasses to protect yourself from potential injuries.

Tools and Materials Needed:
- Sandpaper (various grits)
- Body filler (e.g., Bondo)
- Putty knife or spreader
- Plastic sheeting
- Painter's tape
- Primer
- Touch-up paint (matching your garage door color)
- Paintbrush or small roller
- Clean, lint-free cloths
- Safety gloves and glasses

Step 1: Clean the Damaged Area

Before beginning the repair process, clean the damaged area thoroughly to ensure proper adhesion of the repair materials.

1. Use a clean, lint-free cloth to remove any dirt, debris, or loose paint from the dent or crack and the surrounding area.

2. If necessary, use a mild detergent solution to remove any grease or grime, and allow the area to dry completely.

Step 2: Sand the Damaged Area
Sanding the damaged area creates a rough surface that allows the repair materials to bond more effectively.

1. Use medium-grit sandpaper (e.g., 120-grit) to gently sand the dent or crack and the surrounding area, removing any raised edges or loose material.
2. Feather the edges of the damaged area to create a smooth transition between the repair and the undamaged panel surface.
3. Wipe away any sanding dust with a clean, lint-free cloth.

Step 3: Apply Body Filler
Body filler, such as Bondo, is used to fill in the dent or crack and create a smooth, even surface.

1. Mix the body filler according to the manufacturer's instructions, typically combining a small amount of the filler with a hardening agent.
2. Use a putty knife or spreader to apply the filler to the damaged area, overfilling it slightly to allow for sanding.
3. Allow the filler to cure completely, following the manufacturer's guidelines for drying time.

Step 4: Sand the Filled Area
Once the body filler has cured, sand it to create a smooth, level surface that blends seamlessly with the surrounding panel.

1. Use a coarse-grit sandpaper (e.g., 80-grit) to initially shape the filled area, removing any excess filler.
2. Progress to medium-grit sandpaper (e.g., 120-grit) to further smooth the surface.

3. Finish with fine-grit sandpaper (e.g., 220-grit) to create a smooth, even surface that is level with the surrounding panel.
4. Wipe away any sanding dust with a clean, lint-free cloth.

Step 5: Prime and Paint the Repaired Area
To protect the repaired area and ensure a seamless appearance, prime and paint the repaired section to match the rest of your garage door.

1. Apply painter's tape and plastic sheeting to the area around the repair to protect the undamaged panel surface from overspray.
2. Apply a coat of primer to the repaired area, following the manufacturer's instructions for application and drying time.
3. Once the primer has dried, apply the touch-up paint to the repaired area, using thin, even coats to build up coverage gradually.
4. Allow the paint to dry completely, then remove the painter's tape and plastic sheeting.

Step 6: Inspect and Maintain the Repair
After completing the repair, inspect the area regularly to ensure that the repair remains intact and the panel continues to function properly.

1. Check the repaired area periodically for any signs of cracking, flaking, or separation from the surrounding panel.
2. If necessary, touch up the paint or reapply filler to maintain the integrity and appearance of the repair.
3. Continue to maintain your garage door according to the manufacturer's recommendations, including regular cleaning, lubrication, and inspections.

Conclusion

Repairing minor dents and cracks in your garage door panels is an important part of maintaining the overall health, appearance, and functionality of your garage door. By following the steps outlined in this guide and using the appropriate materials and techniques, you can effectively repair small imperfections and prevent them from developing into more serious issues.

Remember, if you encounter any damage that you're unsure how to repair or that compromises the safety or structural integrity of your garage door, don't hesitate to contact a professional garage door technician for expert advice and assistance. By addressing panel damage promptly and properly, you can extend the lifespan of your garage door and ensure that it continues to provide reliable, secure, and efficient operation for years to come.

Replacing Individual Panels

Sometimes, the damage to a garage door panel may be too severe to repair effectively, or the cost of repairing the panel may be more than the cost of replacing it entirely. In these cases, replacing the individual damaged panel is often the best course of action to restore the appearance, functionality, and security of your garage door. In this section, we'll provide a detailed, step-by-step guide on replacing individual garage door panels.

Safety Warning: Replacing garage door panels involves working with heavy, awkward materials and can be dangerous if not done correctly. If you are not confident in your ability to replace the panel safely, it's best to contact a professional garage door technician for assistance.

Tools and Materials Needed:
- Replacement garage door panel (matching your door's make, model, and color)
- Winding bars (for doors with torsion springs)
- C-clamps or locking pliers
- Socket wrench set
- Screwdrivers (flathead and Phillips)
- Hammer
- Pliers
- Safety gloves and glasses

Step 1: Prepare the Garage Door
Before beginning the panel replacement process, you'll need to secure the garage door in place and release the tension from the springs.

1. Close the garage door completely and disconnect it from the opener.

2. Place C-clamps or locking pliers on the track just above the bottom roller on each side of the door to prevent it from moving.
3. If your door has torsion springs, use winding bars to release the tension from the springs. If you're not comfortable doing this, contact a professional.

Step 2: Remove the Old Panel
With the door secured and the spring tension released, you can now remove the damaged panel.

1. Locate the hinges that connect the damaged panel to the panels above and below it.
2. Remove the bolts or screws that secure the hinges to the damaged panel. You may need to tap the hinge pins out with a hammer if they're difficult to remove.
3. Carefully lift the damaged panel out of the door, being mindful of any sharp edges or broken pieces.

Step 3: Prepare the New Panel
Before installing the new panel, ensure that it matches the size, style, and color of your existing panels.

1. If necessary, transfer any hardware, such as hinges or handles, from the old panel to the new one.
2. Ensure that the new panel is free from any defects or damage before proceeding with the installation.

Step 4: Install the New Panel
With the new panel prepared, you can now install it in place of the old, damaged panel.

1. Position the new panel in the opening left by the removed panel, aligning it with the hinges on the panels above and below.

2. Secure the hinges to the new panel using the bolts or screws removed earlier. Ensure that the panel is level and aligned with the surrounding panels.

3. Test the movement of the door by manually opening and closing it to ensure that the new panel is properly installed and moves smoothly.

Step 5: Restore Spring Tension and Test the Door

After installing the new panel, restore the tension to the springs and test the door's operation.

1. If your door has torsion springs, use the winding bars to restore the proper tension according to the manufacturer's specifications.
2. Remove the C-clamps or locking pliers from the tracks.
3. Reconnect the garage door opener and test the door's operation, ensuring that it opens and closes smoothly and evenly.

Step 6: Maintain the New Panel

To keep your new garage door panel in good condition and prevent future damage, perform regular maintenance and inspections.

1. Clean the panel regularly to remove dirt, grime, and salt (if applicable) that can lead to corrosion or degradation.
2. Inspect the panel periodically for any signs of damage, wear, or deterioration, and address any issues promptly.
3. Follow the manufacturer's recommendations for maintaining and lubricating the hinges, rollers, and other hardware to ensure smooth operation and extend the life of your garage door.

Conclusion

Replacing an individual garage door panel can be a complex and potentially dangerous task, but with the right tools, materials, and guidance, it is possible for a skilled DIY enthusiast to complete the job successfully. By following the steps outlined in this guide and prioritizing safety throughout the process, you can restore the appearance, functionality, and security of your garage door.

However, if you are unsure about any aspect of the panel replacement process or encounter any difficulties along the way, it's always best to contact a professional garage door technician for expert assistance. They have the knowledge, skills, and tools necessary to replace your damaged panel safely and efficiently, ensuring that your garage door continues to provide reliable, secure, and smooth operation for years to come.

Painting and Weatherproofing Panels

Painting and weatherproofing your garage door panels is an important part of maintaining their appearance, protecting them from the elements, and extending their lifespan. Not only does a fresh coat of paint enhance your home's curb appeal, but it also helps to prevent rust, corrosion, and other types of damage that can compromise the integrity and functionality of your garage door. In this section, we'll provide a detailed guide on painting and weatherproofing your garage door panels.

Tools and Materials Needed:
- Cleaning supplies (mild detergent, sponge, bucket)
- Sandpaper (various grits)
- Painter's tape
- Drop cloth or plastic sheeting
- Primer (rust-inhibiting for metal doors)
- Exterior paint (suitable for your door's material)
- Paintbrushes, rollers, or sprayer
- Safety gloves and glasses

Step 1: Clean the Panels
Before painting or weatherproofing your garage door panels, it's essential to clean them thoroughly to ensure proper adhesion of the paint and to remove any dirt, grime, or contaminants that could affect the final result.

1. Mix a mild detergent with warm water in a bucket.
2. Using a sponge, wash each panel thoroughly, paying attention to any particularly dirty or stained areas.
3. Rinse the panels with clean water and allow them to dry completely.

Step 2: Repair Any Damage

If your garage door panels have any dents, cracks, or other damage, repair them before proceeding with painting or weatherproofing.

1. Follow the steps outlined in the "Repairing Minor Dents and Cracks" section to address any minor damage.
2. For more severe damage, consider replacing the affected panel(s) as described in the "Replacing Individual Panels" section.

Step 3: Sand and Prepare the Panels

Sanding the panels creates a smooth surface that allows the primer and paint to adhere more effectively, resulting in a more durable and even finish.

1. Use medium-grit sandpaper (e.g., 120-grit) to lightly sand each panel, focusing on any rough or uneven areas.
2. Wipe away any sanding dust with a clean, lint-free cloth.
3. Apply painter's tape to any areas you don't want to be painted, such as window frames or weatherstripping.

Step 4: Apply Primer

Applying a primer to your garage door panels helps to create a uniform surface, improves paint adhesion, and increases the durability of the final finish.

1. Choose a primer suitable for your garage door's material (e.g., rust-inhibiting primer for metal doors).
2. Apply the primer evenly to each panel using a paintbrush, roller, or sprayer, following the manufacturer's instructions for application and drying time.
3. Allow the primer to dry completely before proceeding to the next step.

Step 5: Paint the Panels

With the panels cleaned, repaired, sanded, and primed, you can now apply the paint to achieve the desired color and finish.

1. Choose an exterior paint suitable for your garage door's material and compatible with the primer you used.
2. Apply the paint evenly to each panel using a paintbrush, roller, or sprayer, following the manufacturer's instructions for application and drying time.
3. Apply additional coats as needed to achieve the desired coverage and color, allowing each coat to dry completely before applying the next.

Step 6: Apply Weatherproofing (Optional)

For added protection against the elements, you may choose to apply a weatherproofing sealant to your garage door panels after painting.

1. Select a weatherproofing sealant compatible with the paint you used and suitable for your door's material.
2. Apply the sealant evenly to each panel, following the manufacturer's instructions for application and drying time.
3. Allow the sealant to dry and cure completely before using the garage door.

Step 7: Maintain the Painted and Weatherproofed Panels

To keep your painted and weatherproofed garage door panels looking and functioning their best, perform regular maintenance and inspections.

1. Clean the panels periodically to remove dirt, grime, and other contaminants that can degrade the finish over time.
2. Inspect the panels regularly for any signs of chipping, peeling, or cracking in the paint or weatherproofing, and touch up as needed.

3. Follow the manufacturer's recommendations for the proper care and maintenance of your specific garage door material and finish.

Conclusion

Painting and weatherproofing your garage door panels is a valuable investment in the appearance, protection, and longevity of your garage door. By following the steps outlined in this guide and using high-quality, suitable materials, you can achieve a beautiful, durable finish that will enhance your home's curb appeal and protect your garage door from the elements for years to come.

Remember, if you're unsure about any aspect of the painting or weatherproofing process, or if you encounter any challenges along the way, don't hesitate to consult with a professional garage door technician or a painting contractor for expert advice and assistance. By maintaining your garage door panels properly and addressing any issues promptly, you can ensure that your garage door continues to provide reliable, secure, and efficient operation while maintaining its attractive appearance.

Insulating Garage Door Panels for Energy Efficiency

Insulating your garage door panels is an effective way to improve your home's energy efficiency, reduce energy costs, and create a more comfortable environment in your garage. Insulated garage doors help to maintain stable temperatures, minimize noise transmission, and prevent cold or heat from entering your living spaces. In this section, we'll provide a comprehensive guide on insulating your garage door panels for enhanced energy efficiency.

Tools and Materials Needed:
- Insulation kit (foam board, reflective insulation, or batt insulation)
- Retainer pins or double-sided tape
- Utility knife
- Measuring tape
- Straight edge or T-square
- Safety gloves and glasses

Step 1: Choose the Right Insulation Material

There are several types of insulation materials suitable for garage door panels, each with its own benefits and installation methods.

1. Foam board insulation: Rigid panels of polystyrene or polyurethane insulation that provide excellent thermal resistance and are easy to install.
2. Reflective insulation: Foil-faced bubble wrap or foil-faced foam that reflects radiant heat and is well-suited for garages in hot climates.
3. Batt insulation: Fiberglass or rock wool insulation that is pre-cut to fit between door panel frames and provides good thermal and acoustic insulation.

Step 2: Measure and Cut the Insulation

Measure your garage door panels and cut the insulation material to fit snugly within each panel frame.

1. Measure the width and height of each panel frame, subtracting about 1/4 inch from each dimension to allow for easier installation.
2. Using a straight edge or T-square and a utility knife, cut the insulation material to the appropriate sizes.
3. For batt insulation, cut the material about 1 inch wider than the panel frame to ensure a snug fit.

Step 3: Install the Insulation

The installation process will vary depending on the type of insulation material you have chosen.

Foam Board Insulation:
1. Clean the inside surface of each garage door panel to ensure proper adhesion.
2. Apply double-sided tape or adhesive to the perimeter of each panel frame.
3. Press the cut foam board insulation into place within each frame, ensuring a snug fit.
4. Use retainer pins or additional adhesive, if necessary, to secure the insulation in place.

Reflective Insulation:
1. Clean the inside surface of each garage door panel.
2. Apply double-sided tape to the perimeter of each panel frame.
3. Cut the reflective insulation to size, allowing for a slight overlap to create a vapor barrier.
4. Press the insulation into place within each frame, ensuring a snug fit and sealing any seams with foil tape.

Batt Insulation:
1. Clean the inside surface of each garage door panel.
2. Cut the batt insulation to size, about 1 inch wider than each panel frame.
3. Press the insulation into place within each frame, ensuring a snug fit without compressing the material.
4. Use retainer pins or straps to hold the insulation securely in place.

Step 4: Seal Gaps and Edges
To maximize the energy efficiency of your insulated garage door panels, seal any gaps or edges to prevent air leakage.

1. Use weatherstripping or garage door bottom seals to seal the gaps between the door and the floor.
2. Apply weatherstripping or foam tape to the sides and top of the door to seal gaps between the door and the frame.
3. Use caulk or expanding foam to fill any remaining gaps or cracks around the perimeter of the door.

Step 5: Maintain and Monitor the Insulation
After installing insulation in your garage door panels, regularly maintain and monitor its effectiveness.

1. Periodically inspect the insulation for any signs of damage, shifting, or compression, and make repairs as needed.
2. Monitor the temperature and humidity levels in your garage to ensure that the insulation is performing as expected.
3. Consider installing a garage door insulation kit with a higher R-value if you live in a particularly cold climate or if you notice that your current insulation is not providing sufficient energy savings.

Conclusion

Insulating your garage door panels is a smart investment in your home's energy efficiency, comfort, and overall value. By following the steps outlined in this guide and choosing the right insulation material for your needs, you can effectively reduce energy costs, stabilize temperatures, and create a more pleasant environment in your garage.

Remember, while insulating your garage door panels is a relatively simple DIY project, it's essential to prioritize safety and follow the manufacturer's instructions for your chosen insulation material. If you encounter any challenges or are unsure about the installation process, don't hesitate to consult with a professional garage door technician or insulation contractor for expert guidance and assistance.

By properly installing and maintaining insulation in your garage door panels, you can enjoy the benefits of improved energy efficiency, reduced noise transmission, and enhanced comfort in your garage and adjacent living spaces for years to come.

Chapter 5
Additional Maintenance and Troubleshooting

Maintaining and Replacing Garage Door Rollers

Garage door rollers are essential components that allow your door to move smoothly and quietly along the tracks. These small wheels, typically made of steel, nylon, or plastic, are subjected to constant wear and tear and may require regular maintenance or replacement to ensure the proper functioning of your garage door. In this section, we'll provide a detailed guide on maintaining and replacing garage door rollers.

Tools and Materials Needed:
- Replacement rollers (match the size and material of your existing rollers)
- Socket wrench set
- Adjustable wrench
- Flathead screwdriver
- Needle-nose pliers
- Safety gloves and glasses
- Lubricant (silicone-based or as recommended by the manufacturer)

Step 1: Inspect the Rollers
Regularly inspecting your garage door rollers can help you identify wear and tear, damage, or other issues that may require maintenance or replacement.

1. Look for signs of wear, such as cracks, chips, or flat spots on the roller surface.
2. Check for any bent or broken roller stems or axles.

3. Ensure that the rollers are properly aligned and rotating freely within the hinges.

4. Listen for any unusual noises, such as squeaking or grinding, which may indicate worn or damaged rollers.

Step 2: Lubricate the Rollers

Regular lubrication of your garage door rollers can help reduce friction, minimize wear, and ensure smooth, quiet operation.

1. Choose a silicone-based lubricant or a lubricant specifically recommended by your garage door manufacturer.

2. Apply a small amount of lubricant to each roller, focusing on the bearing or bushing area where the roller rotates.

3. Wipe away any excess lubricant to prevent dirt and debris from accumulating on the rollers.

4. Open and close the garage door a few times to distribute the lubricant evenly.

Step 3: Replace Worn or Damaged Rollers

If your garage door rollers are excessively worn, damaged, or no longer function properly, they should be replaced to ensure the safe and efficient operation of your door.

1. Close the garage door and disconnect the power to the opener.

2. Use a C-clamp or locking pliers to secure the door in place on the track, preventing it from moving during the replacement process.

3. Starting with the bottom roller, remove the roller from the hinge by gently prying it out with a flathead screwdriver or needle-nose pliers.

4. Insert the new roller into the hinge, ensuring that it is properly seated and aligned.

5. Repeat the process for each remaining roller, working your way up the door.

6. Once all the rollers are replaced, remove the C-clamp or locking pliers and test the door's operation, ensuring that it moves smoothly and quietly.

Step 4: Maintain Proper Roller Alignment
Misaligned rollers can cause your garage door to bind, stick, or come off the tracks, leading to potential damage or safety hazards.

1. Regularly check the alignment of your garage door rollers, ensuring that they are properly seated within the hinges and tracks.
2. If you notice any misalignment, gently tap the roller back into place using a rubber mallet or adjustable wrench.
3. If the misalignment persists or appears to be caused by a bent hinge or track, contact a professional garage door technician for assistance.

Step 5: Schedule Regular Professional Maintenance
While many aspects of garage door roller maintenance can be performed by homeowners, it's important to schedule regular professional maintenance to ensure the longevity and safe operation of your door.

1. Have a certified garage door technician inspect and service your garage door and its components, including the rollers, at least once a year.
2. During these professional maintenance visits, the technician can identify and address any potential issues, make necessary adjustments, and provide expert advice on maintaining your garage door system.

Conclusion

Maintaining and replacing your garage door rollers is a crucial aspect of ensuring the smooth, safe, and efficient operation of your garage door. By regularly inspecting your rollers, providing proper lubrication, replacing worn or damaged components, and maintaining proper alignment, you can extend the life of your garage door system and prevent potential issues.

Remember, while many roller maintenance tasks can be performed by homeowners, it's essential to prioritize safety and consult with a professional garage door technician if you encounter any challenges or are unsure about the proper maintenance procedures. By combining DIY maintenance with regular professional service, you can keep your garage door rollers in top condition and enjoy reliable, quiet operation for years to come.

Adjusting and Replacing Garage Door Cables

Garage door cables are critical components that work in conjunction with the springs to lift and lower your door. These strong, galvanized steel cables are responsible for supporting the door's weight and ensuring that it moves smoothly and evenly. Over time, cables can become worn, frayed, or stretched, requiring adjustment or replacement to maintain the safe and proper operation of your garage door. In this section, we'll provide a comprehensive guide on adjusting and replacing garage door cables.

Safety Warning: Adjusting or replacing garage door cables can be dangerous due to the high tension of the springs. If you are unsure about your ability to safely perform these tasks, it is strongly recommended that you contact a professional garage door technician for assistance.

Tools and Materials Needed:
- Replacement cables (if necessary)
- Adjustable wrench
- Socket wrench set
- Locking pliers or C-clamps
- Safety gloves and glasses
- Ladder

Step 1: Inspect the Cables
Regularly inspecting your garage door cables can help you identify wear, damage, or other issues that may require adjustment or replacement.

1. Look for signs of wear, such as fraying, kinking, or unraveling of the cable strands.

2. Check for any rust or corrosion on the cables, which can weaken their integrity.
3. Ensure that the cables are properly wound around the drums and seated in the grooves.
4. Look for any slack or unevenness in the cables when the door is fully closed.

Step 2: Adjust the Cable Tension
If your garage door cables have become stretched or loose, adjusting their tension can help restore proper operation and balance.

1. Close the garage door and disconnect the power to the opener.
2. Use locking pliers or C-clamps to secure the door in place on the track, preventing it from moving during the adjustment process.
3. Locate the cable drum at the end of the torsion spring (for torsion spring systems) or the pulleys at the top of the door (for extension spring systems).
4. Loosen the set screws on the cable drum or pulleys using a socket wrench.
5. Adjust the cable tension by turning the drum or pulley to wind or unwind the cable. Turn the drum or pulley until the cable is taut and the door is properly balanced.
6. Re-tighten the set screws to lock the cable drum or pulleys in place.
7. Remove the locking pliers or C-clamps and test the door's operation, ensuring that it lifts and lowers smoothly and evenly.

Step 3: Replace Damaged or Worn Cables
If your garage door cables are severely worn, damaged, or broken, they must be replaced to ensure the safe and proper operation of your door.

1. Close the garage door and disconnect the power to the opener.

2. Use locking pliers or C-clamps to secure the door in place on the track.

3. Release the tension from the springs by carefully unwinding them (for torsion springs) or disconnecting the safety cables (for extension springs). This process can be dangerous and should only be performed by experienced individuals or professionals.

4. Disconnect the old cables from the bottom brackets on the door and the cable drums or pulleys at the top.

5. Thread the new cables through the pulleys (if applicable) and attach them to the cable drums or bottom brackets, ensuring they are properly seated and secured.

6. Reattach the springs and wind them to the proper tension, following the manufacturer's specifications.

7. Remove the locking pliers or C-clamps and test the door's operation, making any necessary adjustments to the cable tension or spring balance.

Step 4: Lubricate the Cables and Pulleys

Regular lubrication of your garage door cables and pulleys can help reduce friction, prevent rust, and extend the life of these components.

1. Apply a small amount of silicone-based lubricant or a lubricant recommended by the manufacturer to the cables, focusing on the areas where they contact the pulleys or drums.

2. Lubricate the pulleys or cable drums, ensuring that the lubricant penetrates the bearings or bushings.

3. Wipe away any excess lubricant to prevent dirt and debris from accumulating on the cables or pulleys.

4. Open and close the door a few times to distribute the lubricant evenly.

Step 5: Schedule Regular Professional Maintenance

While some cable maintenance tasks can be performed by homeowners, it is crucial to schedule regular professional maintenance to ensure the longevity and safe operation of your garage door.

1. Have a certified garage door technician inspect and service your garage door and its components, including the cables, springs, and pulleys, at least once a year.
2. During these professional maintenance visits, the technician can identify and address any potential issues, make necessary adjustments, and provide expert advice on maintaining your garage door system.

Conclusion

Adjusting and replacing garage door cables is a vital aspect of maintaining the safe, smooth, and reliable operation of your garage door. By regularly inspecting your cables, adjusting their tension when necessary, replacing damaged or worn components, and providing proper lubrication, you can extend the life of your garage door system and prevent potential hazards.

However, it is essential to emphasize that working with garage door cables and springs can be extremely dangerous due to the high tension involved. If you are unsure about your ability to safely perform cable maintenance or replacement tasks, it is strongly advised that you seek the assistance of a professional garage door technician. By relying on the expertise of trained professionals and following a regular maintenance schedule, you can ensure that your garage door cables remain in excellent condition and continue to support the safe and efficient operation of your door for years to come.

Troubleshooting Noisy Garage Doors

A noisy garage door can be a nuisance and may indicate underlying issues that require attention. Various factors can contribute to a loud or unusual noise when your garage door is in operation, such as worn rollers, loose hardware, or inadequate lubrication. In this section, we'll provide a detailed guide on troubleshooting noisy garage doors and offer solutions to help reduce or eliminate excessive noise.

Step 1: Identify the Type and Location of the Noise
The first step in troubleshooting a noisy garage door is to identify the specific type of noise and its location. This will help narrow down the potential causes and guide you toward the appropriate solution.

Common types of garage door noises include:
1. Squeaking or squealing: Often caused by friction between metal components, such as hinges or rollers, due to a lack of lubrication or worn-out parts.
2. Grinding or grating: May indicate worn or damaged torsion springs, cables, or bearing plates.
3. Rattling or vibrating: Loose hardware, such as hinges, nuts, or bolts, can cause rattling sounds, while a vibrating noise may suggest an issue with the garage door opener or misaligned tracks.
4. Popping or banging: These sudden, loud noises can indicate a variety of issues, such as broken springs, worn rollers, or damaged panels.

Step 2: Inspect and Tighten Loose Hardware
One of the most common causes of noisy garage doors is loose hardware. Inspect and tighten any loose nuts, bolts, or screws to reduce rattling and ensure smooth operation.

1. Check the hinges that connect the door panels and tighten any loose screws or bolts.
2. Inspect the mounting hardware for the tracks, brackets, and garage door opener, and tighten as needed.
3. Ensure that the rollers are securely fastened to the door panels and that the stems are not loose or worn.

Step 3: Lubricate Moving Parts
Proper lubrication is essential for reducing friction and noise in your garage door system. Use a silicone-based lubricant or a lubricant specifically designed for garage doors to lubricate the following components:

1. Hinges: Apply a small amount of lubricant to each hinge, focusing on the pivot points where the panels connect.
2. Rollers: Lubricate the roller bearings or bushings, as well as the stems, to ensure smooth rotation and minimize noise.
3. Tracks: Apply a thin layer of lubricant along the inside of the tracks, where the rollers make contact. Be sure to wipe away any excess lubricant to prevent dirt and debris from accumulating.
4. Springs: Lubricate the torsion springs or extension springs to reduce friction and prevent rust. Be cautious when working near the springs, as they are under high tension.

Step 4: Replace Worn or Damaged Components
If lubrication and tightening hardware do not resolve the noise issue, it may be necessary to replace worn or damaged components.

1. Rollers: Replace any rollers that are worn, chipped, or have flat spots. Opt for nylon or steel rollers with ball bearings for quieter operation.
2. Hinges: Replace any hinges that are bent, damaged, or excessively worn, as they can cause the door to bind or make noise.

3. Springs: If your garage door has torsion springs or extension springs that are worn, stretched, or damaged, they should be replaced by a professional garage door technician, as this can be a dangerous task.
4. Weatherstripping: Replace any worn or damaged weatherstripping along the bottom or sides of the door to reduce noise caused by gaps or uneven contact.

Step 5: Adjust the Garage Door Opener
If the noise appears to be coming from the garage door opener itself, several adjustments can be made to reduce noise and improve performance.

1. Tighten any loose bolts or screws on the opener unit and ensure that it is securely mounted to the ceiling.
2. Adjust the opener's force and travel settings according to the manufacturer's instructions. Improperly adjusted settings can cause the opener to strain or make unusual noises.
3. Consider installing vibration isolators or noise-reduction kits between the opener and the mounting surface to minimize the transfer of vibrations and noise.

Step 6: Maintain Regular Maintenance and Inspections
To prevent noise issues and ensure the longevity of your garage door system, implement a regular maintenance and inspection schedule.

1. Lubricate all moving parts every six months or as recommended by the manufacturer.
2. Tighten loose hardware and inspect for signs of wear or damage monthly.
3. Have a professional garage door technician perform an annual inspection and tune-up to identify and address any potential issues before they cause significant noise or damage.

Conclusion

Troubleshooting a noisy garage door involves identifying the source and type of noise, tightening loose hardware, lubricating moving parts, replacing worn or damaged components, and adjusting the garage door opener as needed. By following these steps and maintaining a regular maintenance schedule, you can significantly reduce or eliminate excessive noise and ensure the smooth, quiet operation of your garage door.

If you are unsure about your ability to safely perform any of these tasks, or if the noise persists after attempting these troubleshooting steps, it is advisable to contact a professional garage door technician for further assistance. They have the expertise and tools necessary to diagnose and resolve complex noise issues, ensuring the safe and efficient operation of your garage door system.

Fixing a Garage Door That Won't Open or Close

A garage door that won't open or close can be frustrating and may indicate a variety of issues, ranging from minor problems like misaligned sensors to more significant concerns like broken springs or damaged tracks. In this section, we'll provide a comprehensive guide on troubleshooting and fixing a garage door that won't open or close.

Safety Warning: Some garage door repairs, such as those involving springs or cables, can be dangerous and should only be attempted by experienced professionals. If you are unsure about your ability to safely perform any of these tasks, contact a professional garage door technician for assistance.

Step 1: Check the Power Supply
Before attempting any mechanical troubleshooting, ensure that your garage door opener is receiving power.

1. Check that the opener unit is plugged in and that the outlet is functioning properly.
2. Ensure that the circuit breaker or fuse associated with the opener hasn't tripped or blown.
3. Test the outlet by plugging in another device to confirm it is supplying power.

Step 2: Inspect the Remote Control and Wall Switch
If the power supply is not the issue, check the remote control and wall switch for proper operation.

1. Replace the batteries in the remote control and test it again.
2. If the remote works, but the wall switch doesn't, check the wiring connections and replace the switch if necessary.

3. If neither the remote nor the wall switch is working, the issue may lie with the opener unit itself or the safety sensors.

Step 3: Check the Safety Sensors
Misaligned, dirty, or obstructed safety sensors can prevent your garage door from opening or closing properly.

1. Ensure that the sensors are properly aligned, with the sending and receiving sensors directly facing each other.
2. Clean the sensor lenses with a soft, dry cloth to remove any dirt or debris that may be blocking the signal.
3. Check for any objects that may be obstructing the sensors' path and remove them.
4. If the sensors are aligned, clean, and unobstructed, but the door still won't close, the issue may be with the wiring or the sensors themselves. Consult a professional for further assistance.

Step 4: Inspect the Tracks and Rollers
Damaged or misaligned tracks, as well as worn or stuck rollers, can prevent your garage door from opening or closing smoothly.

1. Visually inspect the tracks for any signs of damage, such as dents, bends, or gaps.
2. Check that the tracks are properly aligned and level, and adjust them if necessary.
3. Inspect the rollers for any signs of wear, damage, or obstruction, and replace them if needed.
4. Lubricate the rollers and tracks with a silicone-based lubricant to ensure smooth operation.

Step 5: Check the Springs and Cables
Broken or damaged springs, as well as frayed or loose cables, can cause your garage door to malfunction or fail to open or close.

1. Visually inspect the torsion springs (located above the door) or extension springs (located on either side of the door) for any signs of damage, rust, or wear.
2. Check the cables for any signs of fraying, kinking, or slackness.
3. If you suspect that the springs or cables are damaged, do not attempt to repair them yourself. Contact a professional garage door technician for assistance, as these components are under high tension and can be dangerous to work with.

Step 6: Adjust the Force and Travel Settings

If your garage door opener is straining to lift the door or failing to close it completely, the force and travel settings may need adjustment.

1. Locate the force and travel adjustment screws on your opener unit, which are usually labeled "up" and "down" or "open" and "close."
2. Use a flathead screwdriver to make small, incremental adjustments to the settings, following the manufacturer's instructions.
3. Test the door's operation after each adjustment, ensuring that it opens and closes smoothly without straining or reversing.

Step 7: Maintain Regular Maintenance and Inspections

To prevent future issues and ensure the longevity of your garage door system, implement a regular maintenance and inspection schedule.

1. Lubricate all moving parts, including the rollers, hinges, and tracks, every six months or as recommended by the manufacturer.
2. Tighten loose hardware and inspect for signs of wear or damage monthly.
3. Have a professional garage door technician perform an annual inspection and tune-up to identify and address any potential issues before they cause significant problems.

Conclusion

Troubleshooting and fixing a garage door that won't open or close involves a systematic approach to identifying the root cause of the issue. By checking the power supply, remote control, wall switch, safety sensors, tracks, rollers, springs, and cables, and making necessary adjustments or repairs, you can often resolve the problem and restore your garage door to proper operation.

However, it is crucial to prioritize safety when working on your garage door system. If you are unsure about your ability to safely perform any of these tasks, or if the issue persists after attempting these troubleshooting steps, contact a professional garage door technician for further assistance. They have the expertise and tools necessary to diagnose and resolve complex garage door issues, ensuring the safe and efficient operation of your system.

Weatherstripping and Sealing Your Garage Door

Properly weatherstripping and sealing your garage door is essential for maintaining a comfortable, energy-efficient, and pest-free garage. Weatherstripping helps prevent drafts, moisture, and extreme temperatures from entering your garage, while also keeping out dust, debris, and small animals. In this section, we'll provide a comprehensive guide on weatherstripping and sealing your garage door for optimal performance.

Tools and Materials Needed:
- Garage door weatherstripping (bottom seal, threshold seal, and side and top seals)
- Measuring tape
- Scissors or utility knife
- Screwdriver or drill
- Nails or screws
- Hammer
- Caulking gun
- Exterior-grade caulk
- Cleaning supplies (degreaser, rags, and mild detergent)

Step 1: Clean the Garage Door and Frame
Before installing new weatherstripping or seals, clean the garage door and frame to ensure proper adhesion.

1. Use a degreaser or mild detergent to clean the bottom edge of the door, the side and top edges, and the door frame.
2. Remove any old weatherstripping, sealant, or adhesive residue.
3. Allow the door and frame to dry completely before proceeding.

Step 2: Measure and Cut the Bottom Seal

The bottom seal is a crucial component of garage door weatherstripping, as it helps prevent drafts, moisture, and pests from entering through the gap between the door and the floor.

1. Measure the width of your garage door and add an extra inch for overlap.
2. Cut the new bottom seal to the appropriate length using scissors or a utility knife.
3. If your bottom seal has a track or retainer, attach it to the bottom edge of the door using screws or nails, following the manufacturer's instructions.
4. Insert the rubber or vinyl seal into the track or retainer, ensuring a snug fit.

Step 3: Install the Threshold Seal

A threshold seal is an additional barrier that helps prevent drafts and moisture from entering your garage at the bottom of the door.

1. Clean the floor where the threshold seal will be installed, removing any dirt, debris, or old sealant.
2. Measure the width of your garage door opening and cut the threshold seal to the appropriate length.
3. Apply exterior-grade caulk to the underside of the threshold seal, following the manufacturer's instructions.
4. Place the threshold seal on the floor, centered beneath the door, and press firmly to ensure proper adhesion.
5. Allow the caulk to cure according to the manufacturer's guidelines before using the door.

Step 4: Install Side and Top Seals

Side and top seals help prevent drafts and moisture from entering through the gaps between the door and the frame.

1. Measure the height and width of your garage door frame, and cut the side and top seals to the appropriate lengths.
2. If your side and top seals have a track or retainer, attach them to the side and top edges of the door frame using screws or nails, following the manufacturer's instructions.
3. Insert the rubber or vinyl seals into the tracks or retainers, ensuring a snug fit.
4. For adhesive-backed seals, clean the side and top edges of the door frame, then peel off the backing and press the seals firmly into place.

Step 5: Test and Adjust the Weatherstripping

After installing the weatherstripping and seals, test the garage door's operation and make any necessary adjustments.

1. Open and close the garage door several times to ensure that the weatherstripping and seals are not interfering with its operation.
2. Check for any gaps or areas where the weatherstripping or seals are not making proper contact with the door or frame, and adjust as needed.
3. If you notice any excessive resistance or friction when operating the door, trim or adjust the weatherstripping or seals accordingly.

Step 6: Maintain the Weatherstripping and Seals

To ensure the longevity and effectiveness of your garage door weatherstripping and seals, perform regular maintenance and inspections.

1. Periodically clean the weatherstripping and seals with a mild detergent to remove dirt, debris, and any buildup that may interfere with their performance.
2. Inspect the weatherstripping and seals for signs of wear, damage, or gaps, and replace them as necessary.

3. Lubricate any moving parts of the weatherstripping or seals, such as tracks or retainers, with a silicone-based lubricant to ensure smooth operation.

Conclusion

Weatherstripping and sealing your garage door is a simple and effective way to improve energy efficiency, comfort, and pest control in your garage. By following the steps outlined in this guide and using high-quality materials, you can create a tight seal around your garage door that will help keep out drafts, moisture, and unwanted pests.

Remember to clean the door and frame thoroughly before installation, measure and cut the weatherstripping and seals accurately, and test the door's operation after installation to ensure proper functioning. Regular maintenance and inspections will help prolong the life of your weatherstripping and seals, ensuring optimal performance for years to come.

If you encounter any issues or are unsure about your ability to properly install or maintain your garage door weatherstripping and seals, don't hesitate to consult with a professional garage door technician for expert advice and assistance. By prioritizing the weatherproofing of your garage door, you can create a more comfortable, energy-efficient, and secure garage environment.

Conclusion

Throughout this comprehensive guide, we've explored the various aspects of garage door repair, maintenance, and optimization. From understanding the intricate components of your garage door system to troubleshooting common issues and implementing energy-efficient solutions, the knowledge and skills you've gained will prove invaluable in ensuring the longevity, safety, and smooth operation of your garage door.

We've covered the essentials of maintaining and replacing garage door springs, a critical component that plays a vital role in the safe and efficient functioning of your door. You've learned how to identify the signs of worn-out or broken springs, measure and order replacement springs, and follow step-by-step guides for replacing both torsion and extension springs.

We've also delved into the world of garage door openers, exploring the various types available, their features, and how to choose the best opener for your specific needs. With the knowledge gained from this guide, you'll be well-equipped to install, program, and troubleshoot your garage door opener, ensuring convenient and reliable access to your garage.

The importance of properly aligned and maintained garage door tracks cannot be overstated. By understanding the components of your track system and learning how to identify and address issues such as misalignment, damage, or worn-out parts, you can prevent costly repairs and ensure the smooth, quiet operation of your garage door.

We've also addressed the often-overlooked aspects of garage door maintenance, such as weatherstripping and insulation. By implementing effective weatherstripping and sealing techniques, you can improve the energy efficiency of your garage, reduce drafts, and keep out moisture and pests. Additionally, insulating your garage door panels can lead to significant energy savings and a more comfortable garage environment.

While this guide has aimed to provide you with the knowledge and tools necessary to tackle many garage door repair and maintenance tasks on your own, it's essential to recognize the importance of knowing when to call in a professional. For complex repairs or situations where you feel unsure about proceeding safely, don't hesitate to contact a qualified garage door technician who can provide expert assistance and ensure the job is done correctly.

Remember, your garage door is more than just a functional entry point to your home; it's also a significant investment and a critical component of your home's security and curb appeal. By staying proactive in your maintenance efforts, addressing issues promptly, and implementing the strategies and techniques outlined in this guide, you can extend the life of your garage door system and enjoy the many benefits of a well-maintained, properly functioning garage door for years to come.

Thank you for joining us on this journey through the world of garage door repair and maintenance. May the knowledge and skills you've acquired serve you well in keeping your garage door running smoothly, safely, and efficiently. Happy DIY-ing!

www.ingramcontent.com/pod-product-compliance
Lightning Source LLC
Chambersburg PA
CBHW071522220526
45472CB00003B/1114